Applied Solid Dynamics

To Donal

Applied Solid Dynamics

D. G. Gorman BSc, PhD (Strath)

Lecturer in Mechanical Engineering, Queen Mary College, London

W. Kennedy BSc, PhD (Strath), CEng, MIMechE

Senior Lecturer in Dynamics and Control, University of Strathclyde

Butterworths

London Boston Singapore Sydney Toronto Wellington

First published 1988

© **Butterworth & Co. (Publishers) Ltd, 1988**

British Library Cataloguing in Publication Data

Gorman, D.G.
  Applied solid dynamics
  1. Dynamics
  I. Title   II. Kennedy, W.
  620.1′054   TA352

  ISBN 0-408-02309-0

Library of Congress Cataloging-in-Publication Data

Gorman, D.G.
  Applied solid dynamics/D.G. Gorman, W. Kennedy.
    p. cm.
  Includes index.
  ISBN 0-408-02309-0:
  1. Machinery—Dynamics. 2. Dynamics. I. Kennedy, W. II. Title.
TJ170.G67 1988
620.1′054—dc 19   87-34161

Phototypesetting by Thomson Press (India) Limited, New Delhi
Printed and bound in England by Page Bros. Ltd, Norwich, Norfolk.

Preface

The reader, confronted with the preface of yet another textbook on dynamics, might well be forgiven for asking: is such an addition to the already extensive range of texts on this subject really necessary? In our opinion, the answer is yes.

Our experience, gained over many years teaching both undergraduate and postgraduate courses in applied mechanics, has indicated the importance of conveying to students not only the fundamental concepts governing the motion of particles or bodies, but the way in which these fundamentals are associated with practical systems, which at first appear to bear no relationship to any of the theory that has been imparted. Many contemporary textbooks fail to make the successful transition between theory and practice, oversimplifying either or both of these elements and leaving the student stranded somewhere between.

The aim of this book is to help students bridge the gap between theoretical knowledge and practical application, thereby enabling them to approach specific problems with confidence.

Central to this theme is the relationship between rectilinear and rotational systems. Students may understand the basic principles of dynamics when dealing with purely rectilinear systems, but may have difficulty in relating these principles to rotating systems; this difficulty is compounded many times over when the system being analysed possesses both rectilinear and rotational components of motion. In an attempt to overcome these problems, Chapter 1 formulates the concept of dynamically equivalent systems, the use of which enables even the most complex of systems to be represented by a much simpler model—provided certain important criteria are met. The usefulness of this concept is demonstrated in Chapter 2 in the study of the transmission of power through geared systems. In this chapter, also, the reader is introduced to an innovative vector system for the analysis of epicyclic gear transmission.

The transmission of motion by coplanar link mechanisms is investigated in Chapter 3, which also highlights the importance of the simple reciprocating mechanism in relation to the force analysis of bearings and sliding components, and acts as a precursor to the analysis of more complex multicylinder engines in a later chapter.

Chapter 4 builds upon the knowledge imparted in Chapter 3 by demonstrating the effect of intermittent energy transfer in a reciprocating system, and highlights the need for the use of flywheels to act as energy reservoirs in such systems. Attention is also devoted to the general design of flywheels.

Further work on the transmission of power is studied in depth in Chapter 5 where the friction drive, in the form of belts and clutches, is the means of motion and energy transfer. In addition, the manner of energy dissipation, using frictional brake systems, is rigorously examined.

In Chapters 6 and 7 the problems associated with rotational out-of-balance are investigated. This subject is, perhaps, one of the most important aspects of dynamics likely to confront the practising engineer, since in both rotational and reciprocating machines it can often be the major source of vibration. In Chapter 6 a detailed description of the experimental method for determining out-of-balance forces in rotating systems is presented. In Chapter 7 the out-of-balance frame forces and moments associated with a range of positive displacement engines are investigated and recommendations for minimizing these are suggested.

As a natural extension to the work covered in Chapter 3, Chapter 8 expands general plane motion analysis to cover bodies undergoing general space motion, with obvious application to aerospace problems and the kinematic analysis of three-dimensional robotic motion; in addition, examination is made of the related topic of gyrodynamics.

The last five chapters of the book are concerned with vibration theory and the residual effects of this undesirable phenomenon. To some readers this may appear to be a somewhat excessive concentration on this topic; however, in our opinion the extensive coverage of vibration merely reflects the importance of this subject within a whole range of engineering disciplines, particularly in relation to power generation and transmission systems. Vibration theory is introduced at an elementary level in Chapter 9 with an analysis of a single degree of freedom, mass/elastic system performing rectilinear and angular oscillating motions. Once again extensive use is made of the technique of dynamic equivalence in creating simplified mathematical models. The effects of damping, harmonic forcing, transmissibility and seismic excitation are also assessed.

In Chapter 10, systems possessing two degrees of freedom are considered in the absence of damping and external force, but with the added complication of gearing.

The complexity of the system is increased in Chapter 11 with the introduction of multi degree of freedom systems which relate more closely to the practical vibration problems experienced in structural design. Using a simple two degree of freedom system purely as a vehicle, the student is introduced to some of the principles of matrix analysis of such systems. A central theme of this chapter is modal analysis.

In Chapter 12 vibration analysis is extended to cover distributed mass/stiffness systems as opposed to lumped systems. Commencing with the simplest of all distributed systems, namely the stretched wire, analysis proceeds through extensional and torsional vibration of prismatic bars. The lateral vibration of uniform beams is examined together with the effect that rotational motion has on the vibratory response of such components. In addition to the classical analysis, consideration is also given to approximate methods of solution, particularly those that are energy based.

Chapters 9–12 are concerned essentially with the analysis and prediction of the vibrating response of mass/elastic systems, whether such systems are single or multi degree of freedom in nature or are modelled in terms of lumped or distributed parameters. The aim of the practising engineer (and this is consequently of importance to postgraduates and undergraduates as potential engineers) is to reduce, or if possible eliminate completely, the effects of vibration. Chapter 13 is devoted to highlighting some of the ways in which this may be achieved, both by passive system analysis and also in relation to more recently developed active control technology. It is not intended to be an in-depth analysis of the subject but rather an attempt to draw to the reader's attention the range of options open to the engineer in the field of vibratory control.

In the course of the preparation of this text, we were deeply indebted to Julia Shelton, Denis Mudge and Len Bernstein for their proof reading and valuable comments and criticisms; and to Geoff Hancock for many philosophical debates with one of us. The preparation of the manuscript was undertaken by Yvonne Johnson, Marian Parsons and June Neilson—the standard of the finished product being probably the best compliment to their efforts.

Finally, we would like to express our gratitude to our respective families for their patience, understanding and support over the year it took to complete this text; and to the engineering students of Queen Mary College (University of London) and the University of Strathclyde who were our main motivators in this task.

DGG
WK

Contents

Introduction

1.1 Historical review

Dynamics can be defined as that branch of science dealing with the study of the motion of systems under the action of forces, and thus it contrasts with *statics* which is concerned with stationary systems under the action of forces. Dynamics can be divided into two main branches, namely dynamics of solids and dynamics of fluids. Under the action of shearing forces, however, fluids react differently from solids in that they continue to deform as long as the shear forces are applied. For that reason, solid dynamics and fluid dynamics are normally treated separately. In this text we shall concern ourselves with dynamics of solids and, in particular, some of its applications to modern systems.

As in all subjects, although we are mainly concerned with their application, it is of interest to learn something of the history and the people involved in the development of the subject.

The foundations of dynamics may be said to have been laid down by Descartes, Kepler and Galileo. René Descartes (1596–1650), a French philosopher and mathematician, widely regarded as the founder of modern philosophy, introduced analytical geometry—hence the term 'cartesian coordinate system' with which the reader will be familiar. Johan Kepler (1571–1630), a German astronomer, discovered the three basic laws of planetary motion which were published between 1609 and 1619. Galileo Galilei (1564–1642), an Italian mathematician, astronomer and physicist, discovered the uniform period of the pendulum and demonstrated that different bodies of different weight descended at the same rate. Galileo's studies were, however, seriously hindered as a consequence of his theories contradicting those of Aristotle, therefore leading him into continual conflict with the ecclesiastical Inquisition.

On the basis of the work of these three men, Sir Isaac Newton (1642–1727), an English mathematician, astronomer and physicist, formulated his three Laws (or Axioms) of Motion. Newton related the force that acts on a particle to the momentum change it produced, and both these quantities are vectors. Newton essentially derived his three Laws of Motion so that each Law pertained to the three mutually perpendicular directions in space. These laws form the basis of vectorial mechanics. He acknowledged the impact of the earlier work of Descartes, Kepler and Galileo, when he stated: 'If I have seen a little farther than others it is because I have stood on the shoulders of giants'.

About the same time as Newton was working on his Laws of Motion, a German mathematician, Baron Gottfried Wilhelm von Leibniz (1646–1716) was working in this same field, albeit under quite a different approach. Leibniz related the *vis viva* to 'the work of the force', whereby *vis viva* is twice the kinetic energy and 'the work of the force' is called at the present time the 'work function'. In many cases, the work function is simply the potential energy. Later, however, Joseph Louis Lagrange (1736–1813), a French mathematician, and then Sir William Rowan Hamilton (1805–1865), an Irish mathematician, developed analytical dynamics by regarding Leibniz's ideas as the basis of a principle.

In the twentieth century, Albert Einstein (1879–1955), an American physicist, drew attention to the failure of Newtonian mechanics in extreme situations, namely when speeds close to those of light are involved or for events on a molecular scale; however, Hamiltonian dynamics, when suitably interpreted, can cope with these extreme situations and quantum mechanics, which is by and large attributed to Einstein, also has its foundations in Hamiltonian mechanics. In addition it is interesting to note that the design of modern semiconductors (silicon chips) is, to a large extent, based on the theories of Sir William Rowan Hamilton and Robert Brown (1773–1852), a Scottish botanist whose work on the bombardment of particles by molecules has given rise to the term 'Brownian movement'.

Mechanical engineers are not, in general, concerned with the extreme situations of quantum mechanics, and therefore Newtonian mechanics is more than adequate. Attention will consequently be confined to vectorial or Newtonian dynamics and its application to the design and analysis of mechanisms, vehicles, machines, etc.

1.2   Newton's three Laws of Motion

Because of their importance, we shall quote the original Latin form of the three Laws of Motion as set down by Newton in the first edition of his *Principia* in 1687.

Lex I   Corpus omne perseverare in statu suo quiescendi vel movendi uniformiter in directum, nisi quatenus a viribus impressis cogitur statum illum mutare.

Lex II   Mutationem motus proportionalem esse vi motrici impressae, et fieri secundum lineam rectam qua vis illa imprimitur.

Lex III   Actioni contrariam semper et aequalem esse reactionem: sive corporum duorum actiones in se mutuo semper esse aequales et in partes contrarias dirigi.

In 1729 Andrew Motte translated the *Principia* into English; in 1934 F. Cajori, University of California Press, revised this translation, whereby (allowing for errors in the translation from Latin) Newton's three Laws of Motion are:

Law 1   Every body continues in a state of rest, or of uniform motion in a straight line, unless it is compelled to change that state by forces impressed upon it.

Law 2   The change in motion is proportional to the motive force impressed; and it is made in the direction of the straight line in which the force is impressed.

Law 3   To every action there is always opposed an equal reaction; or, the mutual action of two bodies upon each other are always equal, and directed to contrary parts.

The Second Law forms the basis of most analysis of dynamic systems and is usually presented in the more recognizable form:

$$\mathbf{F} = M\mathbf{a}$$

where **F** is the resultant force vector, **a** is the resultant acceleration vector measured in a non-accelerating frame of reference, and M is the mass of the body. Sometimes the Second Law is expressed as the resultant force being equal to the time rate change of momentum with its change in the direction of the force. Both formulations are, however, equally correct when applied to bodies of constant mass. Although, strictly speaking, this law refers to particle motion, it can also be directly applied to rigid bodies (a system of particles bounded by a closed surface that cannot deform) in cases where the motion is purely rectilinear, i.e. where all particles contained within the body move in parallel straight lines. However, it can also be shown to be directly applicable to rotating solid bodies in the form

$$\mathbf{T} = I\boldsymbol{\alpha}$$

where **T** is the resultant torque vector acting on the body about a point fixed in inertial space, $\boldsymbol{\alpha}$ is the angular acceleration vector of the body about the same point, and I is the mass moment of inertia of the body about the point—often referred to as the **polar mass moment of inertia**.

The First Law is a consequence of the Second Law, since there can be no acceleration when the resultant force is zero and therefore the body will either remain at rest or continue to move with constant velocity.

The Third Law defines the rules regarding action and reaction between connected bodies and as such sets out the guidelines for the construction of 'free body diagrams' to which the Second Law is then applied. By means of a practical example, let us now demonstrate how the Second and Third Laws are applied.

1.2.1 Basic vehicle dynamics problem

Consider the case of a rear wheel drive automobile as shown in Figure 1.1 where the *applied* torque of magnitude T_Q at the rear wheels is the driving torque produced by the engine, and F_d is the magnitude of the *applied* aerodynamic drag force acting on the vehicle. We shall assume at this stage that any energy losses from the system, due to heat dissipation, are negligible.

Figure 1.2 shows the free body diagrams of the car body, wheels and road, neglecting any frictional torque at the wheel bearings and any rolling resistance forces (forces at the wheels due to air pressure variations within the tyres).

By inspecting the directions of the forces and torques on each of the free body diagrams, the reader will note that the values and directions of the *actions and reactions*

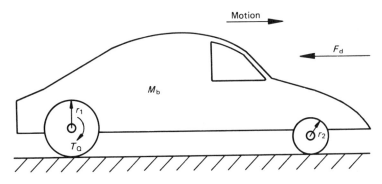

Figure 1.1

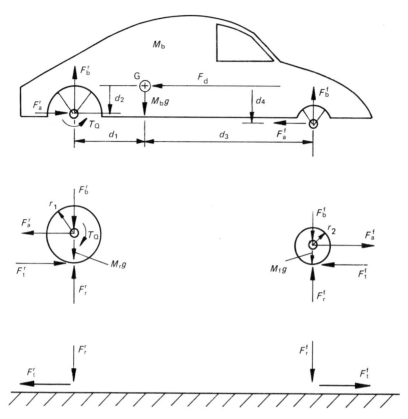

Figure 1.2

are in compliance with the Third Law. Consider now the car body, of mass M_b and centre of mass at the point G, and let us assume that the only motion is the forward rectilinear motion and that the drag force can be taken to act at G. Therefore, from the First and Second Laws, we have

$$F_b^r + F_b^f - M_b g = 0 \quad \text{(vertical equilibrium)}$$
$$d_1 F_b^r - d_2 F_a^r - d_3 F_b^f + d_4 F_a^f - T_Q = 0 \quad \text{(angular equilibrium)}$$
$$F_a^r - F_a^f - F_d = M_b \frac{dv}{dt} \quad \text{(horizontal rectilinear motion)} \tag{1.1}$$

where v is the forward speed of the car body and dv/dt represents the acceleration. Similarly, for the rectilinear motion of the wheels, we have

$$- M_r g + F_r^r - F_b^r = 0 \quad \text{and} \quad - M_f g + F_r^f - F_b^f = 0$$

$$F_t^r - F_a^r = M_r \frac{dv}{dt} \tag{1.2}$$

$$F_a^f - F_t^f = M_f \frac{dv}{dt} \tag{1.3}$$

where M_r and M_f are the mass of the rear and front wheels respectively. Now, for the rotational motion of the wheels,

$$T_Q - r_1 F_t^r = I_r \frac{d\Omega_r}{dt}$$

$$r_2 F_t^f = I_f \frac{d\Omega_f}{dt}$$

where I_r and I_f are the polar mass moments of inertia of the rear and front wheels respectively about their axes of rotation. If at this stage we make the important assumption that no slip occurs at the interface between the wheels and the road, i.e. $\Omega_r = v/r_1$ and $\Omega_f = v/r_2$, then these two latter equations can be rearranged to give

$$T_Q/r_1 - F_t^r = I_r/r_1^2 \frac{dv}{dt} \tag{1.4}$$

and

$$F_t^f = I_f/r_2^2 \frac{dv}{dt} \tag{1.5}$$

Summing equations 1.1, 1.2 and 1.3, we have

$$F_t^r - F_t^f - F_d = (M_b + M_r + M_f) \frac{dv}{dt} \tag{1.6}$$

and if we now make the substitution

$$\frac{dv}{dt} = \frac{ds}{dt} \cdot \frac{dv}{ds} = v \frac{dv}{ds} \tag{1.7}$$

where s is the instantaneous rectilinear displacement of the vehicle, then equation 1.6 can be rewritten as

$$\int (F_t^r - F_t^f - F_d) \, ds = \int (M_b + M_r + M_f) v \, dv \tag{1.8}$$

The left-hand side of equation 1.8 represents the work done by the net resultant force acting on the vehicle, whilst the right-hand side represents the resulting change in kinetic energy associated with the *rectilinear* motion of the vehicle. If we now sum equations 1.4 and 1.5 and make the substitution described by equation 1.7, then

$$\int (T_Q/r_1 - F_t^r + F_t^f) \, ds = \int (I_r/r_1^2 + I_f/r_2^2) v \, dv \tag{1.9}$$

that is, the work done by the net resultant torque acting on the wheels is equal to the change in kinetic energy associated with the *rotational* motion of the wheels.

Equations 1.8 and 1.9 illustrate what is often referred to as the **Principle of Work and Kinetic Energy**, the former as applied to the rectilinear motion only and the latter to the rotational motion only. If we now sum equations 1.8 and 1.9 we have

$$\int (T_Q/r_1 - F_d) \, ds = \int (M_b + M_r + M_f + I_r/r_1^2 + I_f/r_2^2) v \, dv \tag{1.10}$$

Equation 1.10 can be considered as the equation describing the rectilinear motion

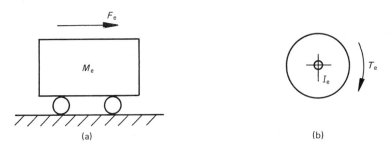

Figure 1.3

of a body of mass M_e acted upon by a single force of magnitude F_e as shown in Figure 1.3a, such that

$$F_e = M_e \frac{dv}{dt} \quad \text{or} \quad \int F_e \, ds = \int M_e v \, dv \tag{1.11}$$

where

M_e = dynamically equivalent mass

$\quad = M_b + M_r + M_f + I_r/r_1^2 + I_f/r_2^2$

F_e = dynamically equivalent force

$\quad = T_Q/r_1 - F_d$

Now, since equations 1.8 and 1.9 describe the Principle of Work and Kinetic Energy with respect only to the rectilinear and rotational components of the motion respectively, equation 1.10 must describe the same principle with respect to the combined motion of the vehicle under the action of the *applied* force F_d and torque T_Q. This can be seen to be the case when one considers that the applied torque of magnitude T_Q acts in the same direction as the rotational motion of the wheels, hence the work term is $+\int T_Q/r_1 \, ds$, whereas the applied drag force of magnitude F_d acts in a direction *opposing* the rectilinear motion of the vehicle, hence the work term, $-\int F_d \, ds$.

Alternatively, reconsider equation 1.10 with $\Omega_r r_1$ replacing v and $\theta_r r_1$ replacing s, where θ_r is the angular displacement of the rear wheels; then after substitution and rearranging we have

$$\int (T_Q - r_1 F_d) \, d\theta_r = \int [(M_b + M_r + M_f)r_1^2 + I_r + I_f r_1^2/r_2^2]\Omega_r \, d\Omega_r \tag{1.12}$$

i.e. we have re-expressed the system as that of a dynamically equivalent torsional system *at the rear wheels* having a polar mass moment of inertia of I_e, and acted upon by a dynamically equivalent torque, T_e, as shown in Figure 1.3b, such that

$I_e = (M_b + M_r + M_f)r_1^2 + I_r + I_f r_1^2/r_2^2$

$T_e = T_Q - r_1 F_d$

In technical publications dealing with vehicle dynamics, the terms **dynamically equivalent mass** (or **inertia**) and **dynamically equivalent force** (or **torque**) are sometimes used, the values of which are calculated from the Principle of Work and Kinetic Energy as applied to the complete system; i.e. F_e or T_e is calculated from the work terms associated with these applied actions on the system, and M_e or I_e is calculated from the total change in kinetic energy of the system. It is important, however, to recall that this dynamic equivalence is only valid when the *zero slip assumption is satisfied*; i.e. there exists a constant relationship between the rotational and rectilinear motion, and heat energy losses are negligible. Assuming that these assumptions are valid, the values of dynamically equivalent mass (or inertia) and force (or torque) associated with a particular system can be readily calculated by means of reference to the Principle of Work and Kinetic Energy as we shall see in the following example.

1.2.2 Basic linear vibration problem (single degree of freedom)

Another type of dynamic situation which lends itself to the concept of dynamic equivalence is the linear vibration analysis of single degree of freedom systems, such as the spring/mass lever system shown in Figure 1.4. As shown, a body of mass M_b and polar mass moment of inertia I_b, about the axis normal to the plane of the diagram and passing through its centre of mass, is rigidly fixed to the lever of mass M (centre of mass at the point G) and polar mass moment of inertia I, about an axis normal to the plane of the diagram and passing through its frictionless pivot point O.

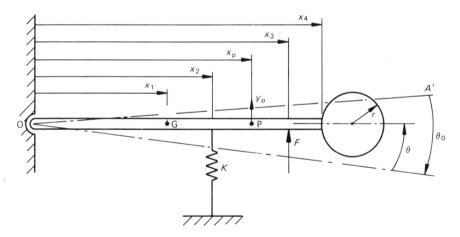

Figure 1.4

In Figure 1.4, the line OA' represents the inclination of the lever at the point where the reaction force on the spring is zero, i.e. the spring is in its normal unloaded condition. The angle θ_0 shown is the angle by which the lever then rotates until it reaches a position of static equilibrium under the action of the gravitational forces and the spring force only, i.e. in the absence of the applied force F. Therefore for this position of the lever, and assuming the angle θ_0 to be small, by taking moments about O we have, from Newton's First Law,

$$- Mgx_1 - M_bg(x_4 + r) + Kx_2^2\theta_0 = 0 \tag{1.13}$$

where K is the spring stiffness. Subsequently, θ is the instantaneous angle, measured from this equilibrium position, which the lever will assume due to the action of the force of magnitude F at the distance x_3 from O as shown.

Let us first derive the equation of motion for the system in the normal manner, i.e. application of the Second Law to the free body diagrams, and then compare it to the equation derived by means of reference to the Principle of Work and Kinetic Energy.

Figure 1.5 shows the free body diagrams for the bodies which constitute this system and the reader should note that the action and reactions are in accordance with Newton's Third Law of Motion. Note also that, since we are assuming horizontal displacements to be negligible, horizontal actions and reactions need not be considered.

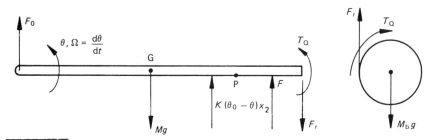

Figure 1.5

Consider the angular motion of the lever about O, whereupon applying Newton's Second Law we have

$$- Mgx_1 + Kx_2^2(\theta_0 - \theta) + Fx_3 - F_r x_4 + T_Q = I\frac{d\Omega}{dt} \tag{1.14}$$

and for the vertical rectilinear motion of the mass M_b

$$F_r - M_b g = M_b \frac{d^2 y}{dt^2}$$

where y is the vertical displacement due to θ, i.e. $y = (x_4 + r)\theta$. Therefore substituting for y and multiplying throughout by $(x_4 + r)$, the above equation becomes

$$F_r(x_4 + r) - M_b g(x_4 + r) = M_b(x_4 + r)^2 \frac{d\Omega}{dt} \tag{1.15}$$

For the rotational motion of the mass M_b about its centre of mass,

$$- T_Q - F_r r = I_b \frac{d\Omega}{dt} \tag{1.16}$$

Subsequently, summing equations 1.14–1.16 and applying the condition described by equation 1.13, we have

$$Fx_3 - Kx_2^2\theta = [I + I_b + M_b(x_4 + r)^2]\frac{d\Omega}{dt} \tag{1.17}$$

Thus, from equation 1.17, the system can be represented as a dynamically equivalent

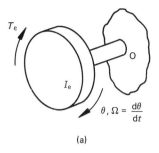

(a)

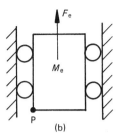

(b)

Figure 1.6

torsional system at O as shown in Figure 1.6a having a polar mass moment of inertia, I_e, equal to $I + I_b + M_b(x_4 + r)^2$, acted upon by a torque, T_e, equal to $Fx_3 - Kx_2^2\theta$.

Alternatively, let us now re-express the actual system as a dynamically equivalent mass at the general point P undergoing rectilinear motion only under the action of a dynamically equivalent force. Let the vertical displacement at P due to the rotational displacement be y_P such that $\theta = y_P/x_P$. Therefore substituting for θ in equation 1.17 and then dividing throughout by x_P we have (remembering that $\Omega = d\theta/dt$)

$$\frac{Fx_3}{x_P} - K\left(\frac{x_2}{x_P}\right)^2 y_P = \left[\frac{I}{x_P^2} + \frac{M_b(x_4 + r)^2}{x_P^2} + \frac{I_b}{x_P^2}\right]\frac{d^2 y_P}{dt^2} \tag{1.18}$$

Therefore the actual system can alternatively be re-expressed as a dynamically equivalent mass, M_e, of value

$$\frac{I}{x_P^2} + \frac{M_b(x_4 + r)^2}{x_P^2} + \frac{I_b}{x_P^2}$$

acted upon by a dynamically equivalent force, F_e, of value

$$\frac{Fx_3}{x_P} - K\left(\frac{x_2}{x_P}\right)^2 y_P$$

undergoing rectilinear motion as described in Figure 1.6b. In this figure, it should be remembered that gravitational effects have been accounted for.

Let us now calculate values for T_e, I_e, F_e and M_e describing the actual system by means of reference to the Principle of Work and Kinetic Energy, which implies that, for equal displacements of the actual and dynamically equivalent systems,

(1)   the total work associated with the applied actions on the actual system must equal the work associated with the dynamically equivalent actions, and
(2)   the change in kinetic energy of both systems must be the same.

Firstly, we shall represent the actual system (Figure 1.4) by the dynamically equivalent torsional system at O as shown in Figure 1.6a. Now, starting from the position shown in Figure 1.4, let us say that when the lever undergoes a small anticlockwise angular displacement, $\delta\theta$, the magnitude of the angular velocity changes from Ω_1 to Ω_2. Therefore, from statement 1, with the product $\delta\theta\,\delta\theta$ being neglected,

$$[Fx_3 - Mgx_1 + Kx_2^2(\theta_0 - \theta) - M_b g(x_4 + r)]\delta\theta = T_e\,\delta\theta \tag{1.19}$$

noting that the moments due to gravity act in a direction opposing the motion; hence the works associated with these forces are negative. Now, substituting the relationship described by equation 1.13, we are left with

$$T_e = Fx_3 - Kx_2^2\theta \qquad \text{(as before)}$$

Consider now statement 2 above. The change in kinetic energy of the actual system is $\frac{1}{2}I_0(\Omega_2^2 - \Omega_1^2)$, where I_0 is the polar mass moment of inertia of the complete system about 0, i.e. (remembering the parallel axis theorem)

$$I_0 = I + I_b + M_b(x_4 + r)^2$$

Now, since the dynamically equivalent system undergoes the same speed change,

$$I_e = I_0 = I + I_b + M_b(x_4 + r)^2 \qquad \text{(as before)}$$

Let us now, using the same approach, re-express the actual system as a dynamically equivalent mass at P undergoing rectilinear motion only under the action of the dynamically equivalent force, F_e, as shown in Figure 1.6b. We start by assigning to the dynamically equivalent mass, M_e, a small linear displacement, δy_P, and in the course of this small displacement the speed of the mass will change from v_1 to v_2. Similarly, we assign the same displacement, δy_P, to the point P of the actual system and in the course of this displacement the linear speed at P will change from v_1 to v_2. Now, with reference to the actual system, the rectilinear displacement, δy_P, and speeds v_1 and v_2 will cause an angular displacement of $\delta\theta = \delta y_P/x_P$ and angular speeds of $\Omega_1 = v_1/x_P$ and $\Omega_2 = v_2/x_P$ about O. Therefore, substituting for $\delta\theta$ (and $\theta = y_P/x_P$) in the left-hand side of equation 1.19, with equation 1.13 implied, the total work done on the actual system by the applied forces must equal the work done on the equivalent system by F_e over the distance δy_P, i.e.

$$\left[\frac{Fx_3}{x_P} - K\left(\frac{x_2}{x_P}\right)^2 y_P\right]\delta y_P = F_e\,\delta y_P$$

Therefore

$$F_e = F\frac{x_3}{x_P} - K\left(\frac{x_2}{x_P}\right)^2 y_P \qquad \text{(as before)}$$

Similarly, by equating the change in kinetic energy of the actual system to that of the dynamically equivalent system, we find that

$$M_e = \left[\frac{I}{x_P^2} + \frac{M_b(x_4 + r)^2}{x_P^2} + \frac{I_b}{x_P^2}\right] \qquad \text{(as before)}$$

In Chapter 9, which deals with vibration of single degree of freedom systems, we shall find that for the purpose of subsequent response analysis it is often very convenient to re-express some practical system as a simple dynamically equivalent system. The choice of whether to select a torsional or a rectilinear equivalent system will depend on the form of the specific problem. Also, in Chapter 2, which deals with gear systems, we shall find it is convenient to re-express these systems as simple dynamically equivalent torsional systems.

Problems

1.1 A 0.3 m diameter hoisting drum is used to raise a load of 5.337 kN vertically. The drum moment of inertia is 10.5 kg m^2 and the friction in the system is equivalent to 440 N m at the drum shaft. Determine the acceleration of the load if the torque applied to the drum shaft is 1450 N m.

Answer: 1.38 m/s^2.

1.2 A bar of mass 0.5 kg is supported between rollers of diameter 0.3 m as shown in Figure 1.7. Each roller is of mass 0.2 kg and has a radius of gyration of 0.1 m about its axis. Determine the velocity of the bar after it has fallen 0.3 m from rest if the rollers roll without slip.

Answer: 1.96 m/s.

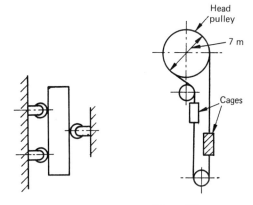

Figure 1.7   **Figure 1.8**

1.3 A motor-driven drum is used to haul trucks up a 1 in 15 slope by cable. The effective drum radius is 0.6 m and its moment of inertia is 470 kg m^2. The motor has a moment of inertia of 8 kg m^2 and runs at 15 times drum speed. Rail traction friction may be taken as 10 N/kN of normal reaction and general rotary friction as 50 N m total at the motor shaft. A 40 kN truck is allowed to accelerate from rest down the incline for 45 s, and then a brake on the motor shaft is applied to stop the truck in a distance of 10 m. Calculate:

(a) the speeds of the truck and the motor after the free run, and
(b) the brake torque used during stopping.

Answer: 15.86 km/h; 1051.95 rev/min; 444 N m.

1.4 The balanced mine hoist shown in Figure 1.8 has a load capacity of 160 kN. The weight of each cage when empty is 56 kN and that of the cable 70 kN. The head pulley is 7 m in diameter and its moment of inertia is 2000 kg m^2. It is powered by a motor having a moment of inertia of $80 \, kg \, m^2$ and a gear reduction between motor and pulley of 20 to 1.

During the hoisting operation, with one cage fully laden and the other empty, the pulley is first accelerated uniformly to a speed of 5 rad/s in 20 s, then revolves at constant speed and finally is uniformly retarded and brought to rest in 20 s.

Calculate the motor torque required during acceleration and during retardation.

Answer: 33.76 kN m; 22.24 kN m.

2

Power transmission through gear systems

2.1 Introduction

The reader will, in some way, be familiar with basic gear systems. The design of gears, and specifically the detail design of gear tooth profiles, will not be covered in any great detail in this text; however, if readers requires such details, they may wish to consult Chapter 12 of *Kinematics and Dynamics of Machines* by G.H. Martin (McGraw-Hill, 1969).

What will be considered, however, given a geared drive system, is how the dynamic characteristics of the system as a whole can be appraised. Prior to commencing this study, a few explanations are necessary with respect to the basic function of two or more meshing gears.

Consider two wheels in peripheral contact as shown in Figure 2.1. If no slip occurs at the instantaneous point of peripheral contact, then

$$\frac{\Omega_b}{\Omega_a} = \frac{\theta_b}{\theta_a} = \frac{D_a}{D_b} = G_{ab}$$

For such a system, unless the coefficient of friction and/or the normal force acting at the peripheral contact point is very high, very little torque could expect to be transmitted due to inevitable slip. The problem of slip may, however, be overcome by introducing 'meshing' gear teeth at the peripheries of both wheels as shown in Figure 2.2 where

p.c.d. = pitch circle diameter

and

circular pitch $= \pi \times$ p.c.d.$/n$

where n is the number of teeth on the wheel. More often, however, the circular pitch is expressed in terms of the **module** (m), such that

$$\text{module} = m = \frac{\text{circular pitch}}{\pi} = \frac{\text{p.c.d.}}{n} \tag{2.1}$$

and must be common for any meshing gear arrangement. Therefore, two meshing gears (of equal module) can be thought of as a system similar to that shown in Figure 2.1, whereby the presence of meshing gear teeth prevent slip from occurring. Furthermore, the diameters D_a and D_b can be replaced by the respective pitch circle

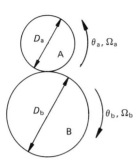

Figure 2.1

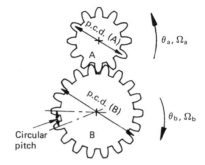

Figure 2.2

diameters, therefore

$$\frac{\Omega_b}{\Omega_a} = \frac{\theta_b}{\theta_a} = \frac{\text{p.c.d.(A)}}{\text{p.c.d.(B)}} \tag{2.2}$$

and substituting from equation 2.1, remembering that module, m, and circular pitch are constant,

$$\frac{\Omega_b}{\Omega_a} = \frac{\theta_b}{\theta_a} = \frac{n_a}{n_b} = \mathbf{G}_{ab} \tag{2.3}$$

where n_a and n_b are the number of gear teeth on wheels A and B respectively. Also, since we have established that no slip can occur, *the ratio Ω_b/Ω_a must be constant irrespective of the torque being transmitted*, and the permissible transmitted torque is limited by the strength of the gear teeth. Let us now proceed to analyse the relationship between the torques applied to a gear system and the ensuing motion.

2.2   General spur gear systems

Consider the general gear system shown in Figure 2.3 where a motor having a polar mass moment of inertia I_m about its axis of rotation provides a torque of magnitude T_Q which drives a machine of polar mass moment of inertia $I_{m/c}$, acted upon by a load torque of magnitude T_L as shown. The transmission between the motor and the machine comprises the three, equal module, gears having polar mass moment of inertia of I_a, I_b and I_c.

Now, if we assume that heat losses from the system are negligible (in modern gearboxes the heat losses constitute only about 3 per cent of the total transmitted power), and because no slip can occur between the gears, we can re-express the system shown as a dynamically equivalent system *at the machine* comprising a dynamically equivalent moment of inertia, I_e, acted upon by a dynamically equivalent torque of magnitude T_e as shown in Figure 2.4. In order to find the values of I_e and T_e we shall apply the Principle of Work and Kinetic Energy as demonstrated in Sections 1.2.1 and 1.2.2. We start by considering a small rotational displacement, $\delta\theta_e$, at I_e of the dynamically equivalent system and the same displacement at the machine of the actual system, as shown in Figures 2.4 and 2.3 respectively. Consequently, the magnitude of the angular displacement at the motor will be $\delta\theta_e \mathbf{G}_{ca}$. Therefore, noting

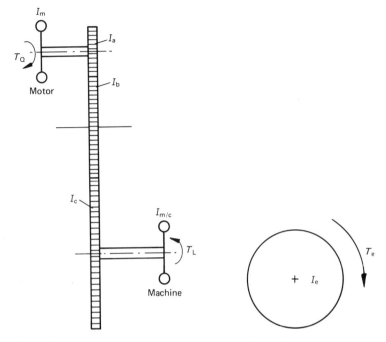

Figure 2.3   **Figure 2.4**

that the sense of the load torque opposes the sense of the motion at the machine
whilst that of the drive torque is in the same sense as the motion at the motor, the
value of T_e can be found from equating the net work term for the actual system to
that of the dynamically equivalent system, assuming that friction at the bearings is
negligible, i.e.

$$T_Q \delta \theta_c \mathbf{G}_{ca} - T_L \delta \theta_c = T_e \delta \theta_c \qquad \text{i.e.} \qquad T_e = T_Q \mathbf{G}_{ca} - T_L \tag{2.4}$$

In the course of this small angular displacement, let the square of the angular speed
of I_e and the machine change of $\delta \Omega_c^2$. Therefore the value of I_e can be found from
equating the change in kinetic of both systems, i.e.

$$\tfrac{1}{2}[(I_m + I_a)\mathbf{G}_{ca}^2 + I_b \mathbf{G}_{cb}^2 + I_c + I_{m/c}]\delta \Omega_c^2 = \tfrac{1}{2} I_e \delta \Omega_c^2$$

i.e.

$$I_e = (I_m + I_a)\mathbf{G}_{ca}^2 + I_b \mathbf{G}_{cb}^2 + I_c + I_{m/c} \tag{2.5}$$

Example 2.1

A motor of moment of inertia 1.05 kg m² drives a machine of moment of inertia
9.5 kg m² through the gear system shown in Figure 2.3, where $n_a = 12$, $n_b = 24$, $n_c = 36$,
$I_a = 0.01$ kg m², $I_b = 0.16$ kg m² and $I_c = 0.81$ kg m².
    The motor supplies a constant torque of 73 N m and the load torque at the machine
is directly proportional to the speed of the machine such that, at a machine speed
of 14.5 rad/s, the load torque is 122 N m.

Find the time taken to accelerate the machine from 14.5 to 25 rad/s and the variation in restoring torque provided by the gearbox mountings during this period, assuming that friction at the bearings and heat losses from the system are negligible.

SOLUTION

$$G_{ca} = \frac{n_c}{n_a} = 3 \quad \text{and} \quad G_{cb} = \frac{n_c}{n_b} = 1.5$$

Now the load torque can be written as $T_L = R\Omega_c$, where R is a constant of proportionality such that, when $\Omega_c = 14.5$ rad/s, $T_L = 122$ N m, thus

$$R = 122/14.5 = 8.414 \text{ N m s/rad}$$

From equation 2.4,

$$T_e = (73 \times 3) - 8.414\,\Omega_c = 219 - 8.414\,\Omega_c \text{ N m}$$

and from equation 2.5

$$I_e = (1.05 + 0.01) \times 3^2 + 0.16 \times (1.5)^2 + (0.81 + 9.5) = 20.21 \text{ kg m}^2$$

Therefore, $T_e = I_e(d\Omega_c/dt)$, i.e.

$$\int_0^{\hat{t}} dt = 20.21 \int_{14.5}^{25} \frac{d\Omega_c}{219 - 8.414\,\Omega_c} = 2.402 \int_{14.5}^{25} \frac{d\Omega_c}{26.03 - \Omega_c}$$

where $\hat{t}$ is the time interval for the machine to increase its speed from 14.5 to 25 rad/s, i.e.

$$\hat{t} = 2.402[\ln(26.03 - \Omega_c)]_{25}^{14.5} = 5.8 \text{ s}$$

Now, when $\Omega_c = 25$ rad/s, $T_L = 8.414 \times 25 = 210.35$ N m, and, when $\Omega_c = 14.5$ rad/s, $T_L = 122$ N m. Furthermore, in this case, Ω_c is in the same sense as Ω_a. Therefore, when $\Omega_c = 25$ rad/s (see Figure 2.5),

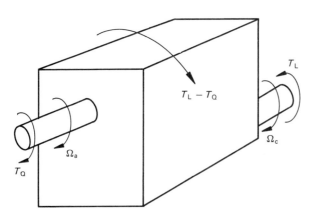

Figure 2.5

Restoring torque provided by gearbox mountings $= T_L - T_Q$
$$= 210.35 - 73$$
$$= 137.35 \, \text{N m}$$

Similarly, when $\Omega_c = 14.5 \, \text{rad/s}$,

$$T_L - T_Q = 122 - 73 = 49 \, \text{N m}$$

Note that if the gearbox mountings were unable to provide this restoring torque, then, by Newton's Second Law, it would be set into a state of angular acceleration.

Example 2.2

Determine the time taken, and the distance covered, for a motor car having the following specifications to attain a speed of 10 m/s from standstill along a flat horizontal road under zero slip conditions (see Figure 2.6a). Assume that friction at all bearings and heat losses from the system are negligible.

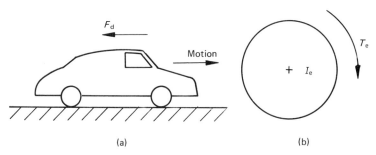

(a)   (b)

Figure 2.6

Specifications:

- For forward speeds between 0 and 10 m/s, the engine torque can be assumed to be of a constant value, $T_Q = 60 \, \text{N m}$.
- The magnitude of the aerodynamic drag force, F_d, opposing motion can be expressed as $F_d = (100 + u^2) \, \text{N}$, where u is the forward speed of the car in m/s.
- Total mass of car, $M_b = 600 \, \text{kg}$.
- Moment of inertia of engine parts, $I_m = 0.25 \, \text{kg m}^2$.
- Total moment of inertia of the road wheels, $I_w = 10 \, \text{kg m}^2$.
- Engine rotational speed/road wheel rotational speed $= 8:1$.
- Radius of road wheels, $r = 0.35 \, \text{m}$.

The moments of inertia of all gearbox and differential components are negligible.

SOLUTION

If Ω_a and Ω_c are the rotational speeds of the engine and road wheels respectively, then

$$G_{ca} = \frac{\Omega_a}{\Omega_c} = 8$$

Once again we start by re-expressing the actual system as a dynamically equivalent torsional system at the wheels (see Figure 2.6b). Consider now a small angular displacement, $\delta\theta$, at the road wheels producing a rectilinear displacement, δs, of the vehicle such that $\delta s = r\delta\theta$ (no slip). Now, if we consider the same small angular displacement of the equivalent system,

$$T_e\,\delta\theta = T_Q\mathbf{G}_{ca}\delta\theta - F_d r\,\delta\theta$$
$$\Rightarrow T_e = T_Q\mathbf{G}_{ca} - F_d r$$
$$= 445 - 0.35u^2$$
$$= 445 - 0.0429\Omega_c^2\,\text{N m}\qquad \text{(since } u = r\Omega_c)$$

Similarly, if the square of the rotational speed of the wheels and the equivalent system both change by $\delta\Omega_c^2$, then

$$\tfrac{1}{2}I_e\delta\Omega_c^2 = \tfrac{1}{2}I_w\delta\Omega_c^2 + \tfrac{1}{2}I_m\mathbf{G}_{ca}^2\delta\Omega_c^2 + \tfrac{1}{2}M_b r^2\delta\Omega_c^2 \qquad (\delta u^2 \times = r^2\delta\Omega_c^2)$$

i.e.

$$I_e = 10 + (0.25 \times 64) + (600 \times 0.1225) = 99.5\,\text{kg m}^2$$

Therefore from $T_e = I_e(d\Omega_c/dt)$, and noting that if $u = 10\,\text{m/s}$ then $\Omega_c = 10/0.35 = 28.57\,\text{rad/s}$, we have

$$\int_0^{\hat{t}} dt = 2319.3 \int_0^{28.57} \frac{d\Omega_c}{101.85^2 - \Omega_c^2}$$

where $\hat{t}$ is the time taken to attain the forward speed of $10\,\text{m/s}$, i.e.

$$\hat{t} = \left(\frac{2319.3}{2 \times 101.85}\right)\left[\ln\left(\frac{101.85 + \Omega_c}{101.85 - \Omega_c}\right)\right]_0^{28.57} = 6.56\,\text{s}$$

To find the distance travelled in this time, we make use of the relationship $T_e = I_e\Omega_c\,d\Omega_c/d\theta$, i.e.

$$\int_0^{\hat{\theta}_c} d\theta = 2319.3 \int_0^{28.57} \frac{\Omega_c\,d\Omega_c}{10373 - \Omega_c^2}$$

where $\hat{\theta}_c$ is the angular displacement of the wheels during this period, i.e.

$$\hat{\theta}_c = (2319.3/2)[\ln(10373 - \Omega_c^2)]_{28.57}^0 = 95\,\text{rad}$$

therefore the rectilinear distance travelled, $s = r\hat{\theta}_c = 33.26\,\text{m}$.

The reader may wish to repeat the solution to this problem by re-expressing the actual system as a dynamically equivalent mass, M_e, undergoing rectilinear motion only by the action a dynamically equivalent force of magnitude F_e.

In the simple gear system which we have so far investigated (Figure 2.3), the speed ratios $\mathbf{G}_{ca}$ and $\mathbf{G}_{cb}$ constituted a very important aspect of the analysis and in this simple case their values were easily calculated. In modern complex gear systems, however, the various speed ratios are not always readily obvious and in such cases it is necessary to apply some form of kinematic analysis in order to obtain the speed ratios. Although there are many such forms of analyses, in this text we shall introduce the Kerr diagram method which was developed and first introduced by Professor W. Kerr of the Royal Technical College (now University of Strathclyde), Glasgow, around 1945 (unpublished). As we shall see in Section 2.3, this is particularly useful for dealing with epicyclic gear systems. However, in the first instance, we shall introduce this method by applying it to simple basic systems.

2.2.1   Kerr diagram for determining speed ratios

Consider the simple gear system shown in Figure 2.7, where A is the input drive shaft, B is an intermediate idler shaft, and C is the output drive shaft. Power is

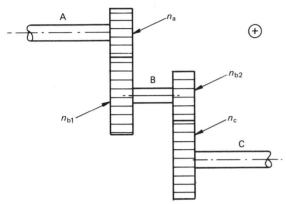

Figure 2.7

transmitted via the gears shown where n_a, n_{b1}, n_{b2} and n_c are the number of teeth on each common module gear. Now, relative to fixed space, denoted by $\oplus$, let us rotate shaft A at a speed of $\Omega_a = X$ rad/s, and represent this by the vector:

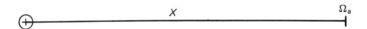

Consequently shaft B will rotate at a speed of $\Omega_b = X(n_a/n_{b1})$ rad/s, in a sense opposite to that of Ω_a with reference to fixed space, thus:

$$\underset{\oplus}{\overset{\Omega_b \qquad X\,(n_a/n_{b1})}{\vdash\!\!-\!\!-\!\!-\!\!-\!\!-\!\!-\!\!-\!\!-\!\!\dashv}}$$

Likewise shaft C will rotate at a speed of $\Omega_c = X(n_a/n_{b1})(n_{b2}/n_c)$ rad/s, opposite in sense to that of Ω_b, i.e.

$$X\left(\frac{n_a}{n_{b1}}\right)\left(\frac{n_{b2}}{n_c}\right)\Omega_c$$

Therefore combining the three vectors we have the complete Kerr diagram:

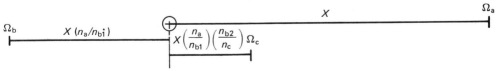

We therefore immediately deduce that

$$\frac{\Omega_a}{\Omega_c} = \left(\frac{n_{b1}}{n_a}\right)\left(\frac{n_c}{n_{b2}}\right) \text{ (in same sense)} = G_{ca}$$

and

$$\frac{\Omega_b}{\Omega_c} = \frac{n_c}{n_{b2}} \text{ (in opposite sense)} = G_{cb}$$

Now consider the system shown in Figure 2.8, where the input drive gear, A, drives the output drive gear C, which is an annular gear with *internally cut teeth*, through an intermediate gear B. For such a system the reader should verify that the complete Kerr diagram is that shown in Figure 2.9.

Therefore, in this case,

$$\frac{\Omega_a}{\Omega_c} = \frac{n_c}{n_a} \text{ (in opposite sense)} = G_{ca}$$

and

$$\frac{\Omega_b}{\Omega_c} = \frac{n_c}{n_b} \text{ (in same sense)} = G_{cb}$$

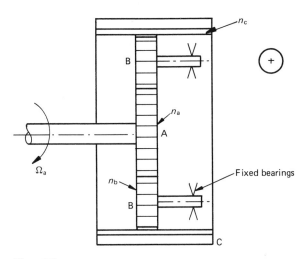

Figure 2.8

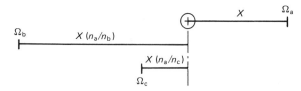

Figure 2.9

As mentioned earlier, this method for determining speed ratios will prove to be extremely useful in Section 2.3 dealing with epicyclic gear systems.

2.3 Epicyclic gear system

Figure 2.10 outlines the arrangement of a basic epicyclic gear system. The input drive is to a **sun wheel**, which, meshing with the **planetary gears** causes these to perform two modes of motion, namely (a) rotation about their own central axis, and (b) rotation around the inner toothed wall of the annular casing, which is usually, but not always, fixed with respect to general space, $\oplus$. This latter motion of the planetary gears, i.e. rotation around the annular casing, is normally arranged as the output drive as shown, and it is this motion that distinguishes the epicyclic gear arrangement from the more basic systems analysed in Section 2.2. The first stage in the analysis of such systems is to establish the various speed ratios of the system. In so doing, if one attempts to construct a Kerr diagram for such a system, then immediately the question is posed: How can the rotation around the annular casing be represented? This can be accommodated as follows:

(1) If the outer annular casing is not free to rotate, free it, i.e. allow it to rotate.
(2) Restrict the motion of the planetary gears in such a manner as to allow them to rotate about their own central axis, but, not to rotate around the inside of the casing. We are thus suppressing the 'epicyclic effect'.

These two steps, in effect, reduce the system to that of a standard gear system similar to that shown in Figure 2.8. Now,

(3) rotate any of the gears (usually the sun wheel), at X rad/s and complete the Kerr

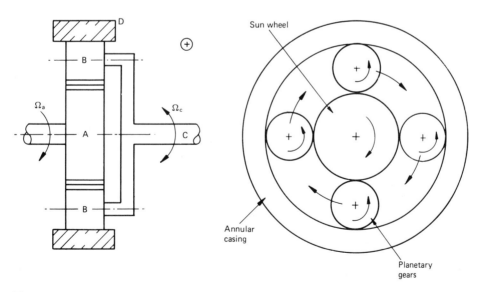

Figure 2.10

diagram as before—note that the fixed space point ⊕ will correspond to the suppressed motion of the planetary gears around the casing—and

(4)   having completed the Kerr diagram, move the fixed space point, ⊕, to its true position, which in this case is the point on the diagram corresponding to the rotation of the casing.

Let us now apply this procedure to the system shown in Figure 2.10, where we will assume that the annular casing, D, is fixed rigid. Therefore, we *free* D, hold shaft C *fixed*, and construct the Kerr diagram with fixed space point ⊕ corresponding to Ω_c. We commence the Kerr diagram by rotating the sun wheel, A, at a rate of X rad/s in any sense, i.e.

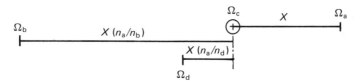

where n_a, n_b and n_d are the number of equal module gear teeth on gears A, B and D respectively. Now apply step 4 by moving the fixed space point ⊕ to the point on the diagram corresponding to Ω_d, i.e.

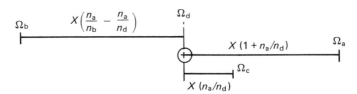

Therefore

$$G_{ca} = \frac{\Omega_a}{\Omega_c} = \frac{n_d}{n_a} + 1 \qquad \text{(acting in the same sense)}$$

$$G_{cb} = \frac{\Omega_b}{\Omega_c} = \frac{n_d}{n_b} - 1 \qquad \text{(acting in the opposite sense)}$$

Having now determined the system speed ratios we can then proceed, as in Section 2.2, to analyse the ensuing motion of these gear systems under the actions of applied torques.

It must, however, be emphasized at this stage that, although epicyclic gear systems may differ in configuration from that shown in Figure 2.10, the existence and described motion of planetary gears are always present. Furthermore, by means of an example at a later stage, we will consider the case where both the annular casing and sun wheel are made to rotate simultaneously.

Example 2.3

A motor of moment of inertia, $I_m = 0.8 \, \text{kg m}^2$, attached to the sun wheel A, drives a machine of moment of inertia, $I_{m/c} = 1.2 \, \text{kg m}^2$ attached to the planetary cage shaft C, through the epicyclic gear system shown in Figure 2.10.

For the system shown, the fixed annulus, D, with 100 internal teeth meshes with the four planetary gears, B, each having 30 teeth and carried by the arm (planetary cage), C. The planetary gears mesh with the sun wheel, A, which has 40 teeth. All the gears are of 1 mm module. Each of the planetary gears has a mass of 0.32 kg and a moment of inertia of 0.036 g m² about their own central polar axis. The sun wheel has a moment of inertia of 0.12 g m² about its own central polar axis. The planetary cage has a moment of inertia of 0.5 g m² about its central axis of rotation.

(a) Determine the speed ratios G_{ca} and G_{cb}.
(b) Determine a dynamically equivalent torque and moment of inertia, referred to the output drive shaft, which could replace the actual system.
(c) If the motor supplies 10 kW of power at 1000 rev/min, calculate the torque the mountings of the annular casing, D, must provide in order to keep the casing stationary at the constant motor speed of 1000 rev/min.
(d) Assuming that the motor speed of 1000 rev/min represents the steady-state speed of the system and
   (i) the motor exhibits constant torque characteristics,
   (ii) the load torque at the machine is proportional to its speed, find the time taken for the speed to reach 95 per cent of the steady-state speed from standstill.

SOLUTION

(a) The speed ratios G_{ca} and G_{cb} are obtained from the Kerr diagram, previously constructed. Therefore

$$G_{ca} = \frac{\Omega_a}{\Omega_c} = \frac{n_d}{n_a} + 1 = 3.5 \qquad \text{(in the same sense)}$$

and

$$G_{cb} = \frac{\Omega_b}{\Omega_c} = \frac{n_d}{n_b} - 1 = 2.333 \qquad \text{(in opposite sense)}$$

(b) The dynamically equivalent torque referred to C, namely T_e, is obtained in the manner demonstrated in Section 2.2, i.e.

$$T_e = T_Q G_{ca} - T_L$$

where T_Q and T_L are the motor and load torques respectively.
  Similarly, the dynamically equivalent moment of inertia, I_e, is

$$I_e = (I_c + I_{m/c}) + 4G_{cb}^2 I_b + 4M_b r^2 + (I_a + I_m)G_{ca}^2$$

where I_a and I_b are the moments of inertia of gears A and B about their own central rotational axes respectively, and I_d is the moment of inertia of the planetary cage about its own central axis. The reader will note the inclusion of the term $4M_b r^2$ in the above expression, where M_b is the mass of a singular planetary gear, B, and r is the radial distance between the central axes of B and D (or A). This term is derived from the kinetic energy of the four planetary gears associated with their motion around the inside of the fixed annulus.
  The value of r can be obtained from the geometry of gears A and B (see Figure 2.3)

where

$$r = \tfrac{1}{2}[\text{p.c.d.(A)} + \text{p.c.d.(B)}]$$

and

$$\text{p.c.d.(A)} = n_a \times \text{module} = 40\,\text{mm}$$
$$\text{p.c.d.(B)} = n_b \times \text{module} = 30\,\text{mm}$$

Hence $r = 35\,\text{mm}$, and substituting in the above expression for I_e gives

$$I_e = 11\,\text{kg}\,\text{m}^2$$

(c) Motor torque at A,

$$T_Q = \frac{10 \times 10^3}{1000 \times \pi/30}$$
$$= 95.49\,\text{N}\,\text{m}$$

and this torque acts in the same sense as Ω_a. Now for constant speed conditions, $T_e = 0$, i.e. $T_L = T_Q G_{ca} = 334.2\,\text{N}\,\text{m}$. Figure 2.11 therefore shows the torques acting

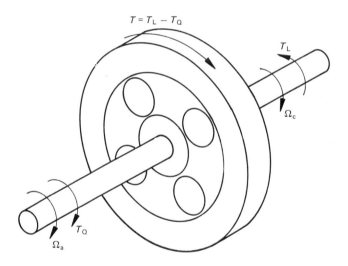

Figure 2.11

on the system. Hence, the torque provided by the casing mountings, T, is

$$T = T_L - T_Q = 334.2 - 95.49 = 238.7\,\text{N}\,\text{m}$$

and acts in the same sense as Ω_a.

(d) If the motor is of constant torque characteristics, then the value of this torque is $95.49\,\text{N}\,\text{m}$ as in (c), and at a motor speed of $1000\,\text{rev/min}$ the load torque at C is $334.2\,\text{N}\,\text{m}$. Now if the load torque, T_L, at the machine is proportional to the speed of the machine, i.e.

$$T_L = R\Omega_c$$

where R is a constant, then

$$R = \frac{334.2}{\Omega_c} \quad \text{at steady speed conditions}$$

$$= \frac{334.2}{\Omega_a/G_{ca}}$$

$$= 11.17\,\text{N m s/rad}$$

Therefore applying Newton's Second Law of Motion to the dynamically equivalent system at C, we have

$$T_e = 95.49\,G_{ca} - 11.17\Omega_c = I_e\frac{d\Omega_c}{dt}$$

and substituting values gives

$$\int_0^{\hat{t}} dt = (11/11.17)\int_0^{28.44}\frac{d\Omega_c}{29.95 - \Omega_c}$$

where $\hat{t}$ is the time taken for shaft C to obtain a speed of 95 per cent of $(1000/G_{ca})$ rev/min from standstill, i.e. 28.44 rad/s. Hence

$$\hat{t} = 0.986[\ln(29.95 - \Omega_c)]_{28.44}^0 = 2.946\,\text{s}$$

The reader may wish to investigate the problem associated with finding the time taken to reach the steady-state speed of 29.92 rad/s at C.

Example 2.4

A schematic section of an epicyclic gear system is shown in Figure 2.12. It comprises a 40-toothed sun wheel, A, on the input drive shaft, meshing with three planetary gears, B, which are carried by a planetary cage connected to the output drive shaft, C. The annular casing, D, having 100 internal gear teeth, is free to rotate and also meshes with the three planetary gears as shown.

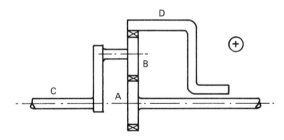

Figure 2.12

If the input drive rotation at the sun wheel is at a speed of 500 rad/s, find the speed of the output drive shaft and its sense relative to the input drive rotation if the annular casing is made to rotate at 250 rad/s in (a) the same sense as the input drive rotation, and (b) the opposite sense to the input drive rotation at A.

SOLUTION

Since all gear teeth are of equal module, the planetary gears (B) must each have $(100 - 40)/2$ teeth, i.e. 30. Once again we start by constructing the Kerr diagram on the basis of C being fixed, and rotating A at X rad/s. Note that, since in this case there is no need to free D, the preliminary Kerr diagram is

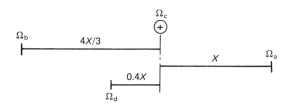

where Ω_d, Ω_a, Ω_b and Ω_c denote the rotational speeds of D, A, B and C respectively.

At this stage we place the fixed space point, $\oplus$, at its appropriate position, which in this case will not be at Ω_d (or Ω_a or Ω_b).

(a) Since Ω_d (250 rad/s) $< \Omega_a$ (500 rad/s) but acts in the same sense, the fixed space point, $\oplus$, must be located thus:

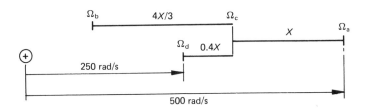

i.e.

$1.4X = (500 - 250)\,\text{rad/s}$

giving

$X = 178.57\,\text{rad/s}$

therefore

$\Omega_c = 250 + 0.4X = 321.43\,\text{rad/s}$

and acts in the same sense as Ω_a.

(b) Once again since $\Omega_d < \Omega_a$, but acts in an opposite sense, the fixed space point, $\oplus$, must be located thus:

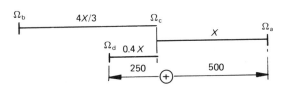

$$1.4X = 750\,\text{rad/s}$$

giving

$$X = 535.714\,\text{rad/s}$$

therefore

$$\Omega_c = 250 - 0.4X = 35.714\,\text{rad/s}$$

and acts in a sense opposite to that of Ω_a.

2.3.1   Automobile 'overdrive' unit

A number of modern automobiles are fitted with an 'overdrive' unit. This consists of an auxiliary, single ratio gearbox which operates on the epicyclic principle. Such a unit, in its operational mode, is shown schematically in Figure 2.13.

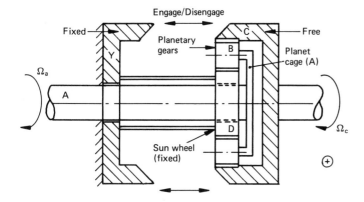

Figure 2.13

In such an arrangement, the sun wheel, D, is fixed and the input drive shaft and planetary cage, A, rotate the planetary gears, B, around the fixed sun wheel and the internally toothed annulus, C, which forms the output drive.

Let us determine the speed ratio, $G_{ac}(=\Omega_c/\Omega_a)$ for this system by constructing the Kerr diagram, initially with the planetary cage fixed and the sun wheel free and rotating at X rad/s, i.e.

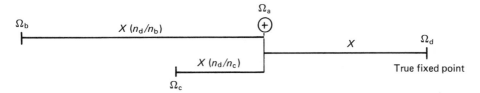

where n_d, n_b, and n_c are the number of teeth on sun wheel, planetary gears and annulus respectively. Now moving the fixed space point, $\oplus$, to Ω_d we deduce that

$$G_{ac} = \frac{\Omega_c}{\Omega_a} = 1 + \frac{n_d}{n_c}$$

and, as such, G_{ac} is always greater than unity. Hence one can deduce why this unit is termed an overdrive unit. When the overdrive unit is disengaged, electro-hydraulically, the following changes occur simultaneously (see Figure 2.13):

(1) the fixed annulus, Y, is freed and locked into the input drive, A, and simultaneously 'locks on' (by friction) to the internally geared annulus, C, which drives the output shaft, and
(2) the sun wheel, D, is freed and allowed to rotate with the annulus, Y, and input drive, A.

The net result of this is to reduce the speed ratio G_{ac} to unity, i.e. the overdrive unit, in effect, reduces to a solid rotating mass.

Problems

2.1 Wheels A, B, C and E have 36, 24, 90 and 30 teeth respectively on the epicyclic gear train shown diagrammatically in Figure 2.14. Calculate the speed of A when

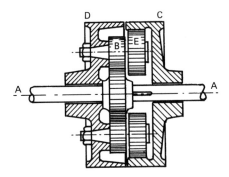

Figure 2.14

C is held and D is driven at 100 rev/min. If the torque on C is limited to 4000 N m, what power will the gear transmit without slip occurring?

Answer: + 300 rev/min; 6.28 kW.

2.2 An epicyclic gear train is shown diagrammatically in Figure 2.15. Wheels B, C and A have 35, 45 and 35 teeth respectively. All teeth are of equal module. D is driven at + 1200 rev/min. Find:

(a) the speed of A when E is held,
(b) the numbers of teeth on G and F for a speed of − 300 rev/min at A when F is held.

Answer: + 533 rev/min; 31; 39.

2.3 Figure 2.16 shows an epicyclic gearbox in which wheels A, B and C have 23,

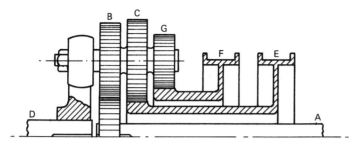

Figure 2.15

33 and 27 teeth respectively. The input shaft A is driven at 320 rev/min, while shaft B is held stationary by a brake.

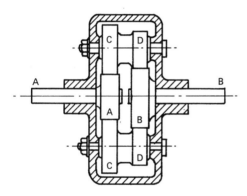

Figure 2.16

Determine the speed of the pulley casing E. If the brake on shaft B is partially released to allow shaft B to run at 100 rev/min in the same direction as shaft A, calculate the new speed of the pulley casing E.

Answer: 250 rev/min in opposite direction to shaft A; 72 rev/min in opposite direction to shaft A.

2.4   A motor of moment of inertia 1.05 kg m² supplies a constant torque of 73.4 N m and drives a machine of moment of inertia 9.5 kg m² through the epicyclic gearbox arrangement, as shown in Figure 2.17.

The resistance torque of the machine is directly proportional to its speed, and the constant of proportionality is 8 N m s/rad. For the epicyclic gear arrangement, the wheel D (40 teeth) is attached to shaft Y which is held stationary; there are two pairs of compound planets, B (30 teeth) and C (50 teeth), rotating freely on pins attached to plate Z, which can rotate independently about axis XY. All teeth are of 6 mm module. B and C together have a combined mass of 3 kg and a polar radius of gyration of 75 mm. A has a polar moment of inertia of 0.906 kg m², and Z, together with attached pins, of 0.35 kg m².

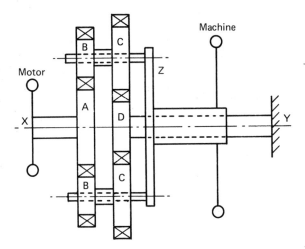

Figure 2.17

Find the time required to increase the speed of the machine from 0 to 5 rad/s.

Answer: 3.3 s.

3

Dynamics of a solid body in general plane motion

In Chapter 1, we considered the application of Newton's Laws of Motion to bodies undergoing rectilinear motion and rotation about a fixed point in a single plane. Although there are a considerable number of applications associated with such a study, there are many more systems which comprise bodies moving in general free space, i.e. simultaneously along the OX, OY and OZ axes of the normal space coordinate system. This form of motion will be examined in Chapter 8. However, as a preliminary to general space motion, let us consider the case, and the applications thereof, where the motion of all particles on a body are confined to the plane surface bounded by the normal axes, OX and OY, as shown in Figure 3.1, i.e. the motion does not have any components along the OZ axis.

Consider the particles, P and Q, on the body shown in Figure 3.1, which, for the present, can be considered to be of a flexible, spongy material. Now, if the motion of the body is such that the particles P and Q (and any other particles within the body) are free to move *independently* of each other, then the body is said to be undergoing *general plane motion*.

3.1 Kinematics of general plane motion

Consider any two particle points P and Q of the body shown in Figure 3.1 undergoing general plane motion. Furthermore we will assume that the displacement, velocity and acceleration vectors of P relative to the fixed point O, i.e. $\boldsymbol{\delta}_{OP}$, $\mathbf{v}_{OP}$ and $\mathbf{a}_{OP}$ respectively, are fully specified. Now consider the instantaneous displacement of Q relative to P, $\boldsymbol{\delta}_{PQ}$, shown vectorially in Figure 3.2, where

$$\boldsymbol{\delta}_{PQ} = \mathbf{i}r\cos\phi + \mathbf{j}r\sin\phi \tag{3.1}$$

and $\mathbf{i}$ and $\mathbf{j}$ are unit vectors along OX and OY respectively. Furthermore, since in general r and ϕ vary with time, the velocity of Q relative to P, $\mathbf{v}_{PQ}$, is

$$\begin{aligned}
\mathbf{v}_{PQ} &= \mathrm{d}(\boldsymbol{\delta}_{PQ})/\mathrm{d}t \\
&= \mathbf{i}(\dot{r}\cos\phi - r\dot{\phi}\sin\phi) + \mathbf{j}(\dot{r}\sin\phi + r\dot{\phi}\cos\phi)
\end{aligned} \tag{3.2}$$

(where $\dot{r} = \mathrm{d}r/\mathrm{d}t$ and $\dot{\phi} = \mathrm{d}\phi/\mathrm{d}t$), or

$$\mathbf{v}_{PQ} = \dot{r}(\mathbf{i}\cos\phi + \mathbf{j}\sin\phi) + r\dot{\phi}(\mathbf{j}\cos\phi - \mathbf{i}\sin\phi) \tag{3.3}$$

From equation 3.3 we observe that $\mathbf{v}_{PQ}$ contains two components:

(1) *Component 1* $= \dot{r}(\mathbf{i}\cos\phi + \mathbf{j}\sin\phi)$. This component, shown diagrammatically in

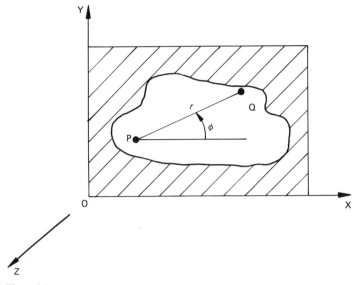

Figure 3.1

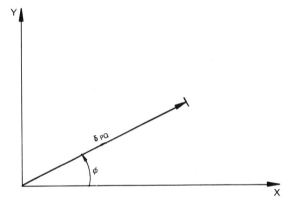

Figure 3.2

Figure 3.3 (where p′ represents the terminal point of the vector $\mathbf{v}_{OP}$), is termed the *radial component* of velocity of Q relative to P, $\mathbf{v}_{PQ}^r$. The magnitude of $\mathbf{v}_{PQ}^r$ is $\dot{r}$ and its line of action is parallel to the line joining the particle points P and Q, and acts in a direction similar to that of the vector $\boldsymbol{\delta}_{PQ}$ (see Figure 3.2) if $\dot{r}$ is positive (increasing), and vice versa.

(2) *Component 2* $= r\dot{\phi}(\mathbf{j}\cos\phi - \mathbf{i}\sin\phi)$. This component, shown diagrammatically in Figure 3.3, is termed the *tangential component* of the velocity of Q relative to P, $\mathbf{v}_{PQ}^t$. The magnitude of $\mathbf{v}_{PQ}^t$ is $r\dot{\phi}$, the line of action is perpendicular to the line joining P and Q, and acts in a direction obtained by rotating the direction of the vector $\boldsymbol{\delta}_{PQ}$ through 90° in the direction of rotation of Q about P.

Hence, from Figure 3.3, the velocity of Q relative to P, $\mathbf{v}_{PQ}$, can be represented

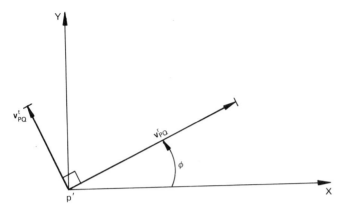

Figure 3.3

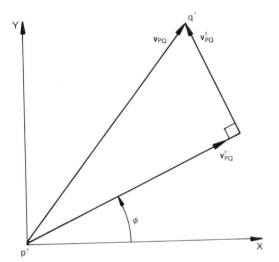

Figure 3.4

graphically as shown in Figure 3.4, where q′ represents the terminal point of this vector.

Let us now turn our attention to the acceleration of Q relative to P, i.e. $\mathbf{a}_{PQ}(= d\mathbf{v}_{PQ}/dt)$, which upon differentiating equation 3.2 gives

$$\mathbf{a}_{PQ} = \ddot{r}(\mathbf{i}\cos\phi + \mathbf{j}\sin\phi) + r\ddot{\phi}(\mathbf{j}\cos\phi - \mathbf{i}\sin\phi)$$
$$- r\dot{\phi}^2(\mathbf{i}\cos\phi + \mathbf{j}\sin\phi) + 2\dot{r}\dot{\phi}(\mathbf{j}\cos\phi - \mathbf{i}\sin\phi) \tag{3.4}$$

where $\ddot{r} = d^2r/dt^2$ and $\ddot{\phi} = d^2\phi/dt^2$.

From equation 3.4 we see that $\mathbf{a}_{PQ}$ contains four distinct components:

(1) *Component 1* $= \ddot{r}(\mathbf{i}\cos\phi + \mathbf{j}\sin\phi)$. This component, shown diagrammatically in Figure 3.5a (where the point p″ denotes the terminal point of the vector $\mathbf{a}_{OP}$), is

termed the **radial component** of the acceleration of Q relative to P, $\mathbf{a}_{PQ}^r$. The magnitude of this component is $\ddot{r}$ and its line of action is parallel to the line joining the particle points, P and Q. The direction of this component is that of the direction of the vector $\boldsymbol{\delta}_{PQ}$ if $\ddot{r}$ is positive, and vice versa.

(2)   *Component 2* $= r\ddot{\phi}(\mathbf{j}\cos\phi - \mathbf{i}\sin\phi)$. This component, shown diagrammatically in Figure 3.5a, is termed the **tangential component** of the acceleration of Q relative to P, $\mathbf{a}_{PQ}^t$, of magnitude $r\ddot{\phi}$. The line of action of this component is perpendicular to the line joining P and Q, and its direction is obtained by rotating the direction of the vector $\boldsymbol{\delta}_{PQ}$ through 90° in the direction of rotation of Q about P if $\ddot{\phi}$ is positive. If $\ddot{\phi}$ is negative, rotate the direction of the vector $\boldsymbol{\delta}_{PQ}$ through 90° in the opposite direction to that of the rotation of Q about P.

(3)   *Component 3* $= -r\dot{\phi}^2(\mathbf{i}\cos\phi + \mathbf{j}\sin\phi)$. This component, shown diagrammatically in Figure 3.5b, is termed the **centripetal component** of the acceleration of Q relative to P, $\mathbf{a}_{PQ}^{ct}$, of magnitude $r\dot{\phi}^2$. The line of action of this component is parallel to the line joining P and Q, and the direction is *always opposite* to that of the direction of the vector $\boldsymbol{\delta}_{PQ}$.

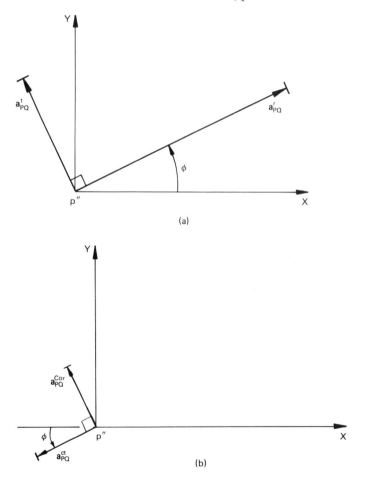

(a)

(b)

Figure 3.5

(4)   *Component 4 = $2\dot{r}\dot{\phi}(\mathbf{j}\cos\phi - \mathbf{i}\sin\phi)$.* This component, shown diagrammatically in Figure 3.5b, is termed the **Coriolis component** of the acceleration of Q relative to P, $\mathbf{a}_{PQ}^{Cor}$ of magnitude $2\dot{r}\dot{\phi}$ (after the French civil engineer Gaspard G. Coriolis (1792–1843)). The line of action of this component is perpendicular to the line joining P and Q and its direction can always be determined by rotating the direction of the vector $\mathbf{v}_{PQ}^{r}$ (see Figure 3.3) through 90° in the direction of rotation of Q about P.

Hence for the case where $\dot{r}, \ddot{r}, \dot{\phi}$ and $\ddot{\phi}$ are all positive, Figure 3.6 shows the complete construction of the acceleration vector of Q relative to P, $\mathbf{a}_{PQ}$, where the point q″ is used to denote the terminal point of this vector.

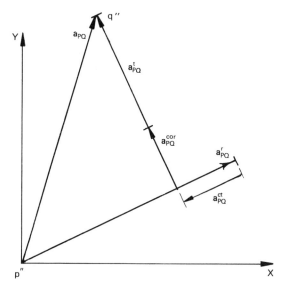

Figure 3.6

Although there are countless applications of such a study, perhaps the most common, with respect to mechanical engineering systems, is the application to *coplanar mechanisms* which form the integral workings of most common machines. As we will see, by means of the following examples, the results of the above study will enable us to construct the complete *velocity and acceleration vector diagrams* for specified cycle stages of these types of mechanisms.

Example 3.1

For the mechanism shown in Figure 3.7a at the instant shown, construct the velocity and acceleration vector diagrams, given that the crank OA rotates in a clockwise sense at a constant speed of 150 rev/min as shown. Hence find the absolute acceleration of the point G and the angular acceleration of the link AGB.

SOLUTION

Since the space points O and C are fixed, all velocity and acceleration vectors relative

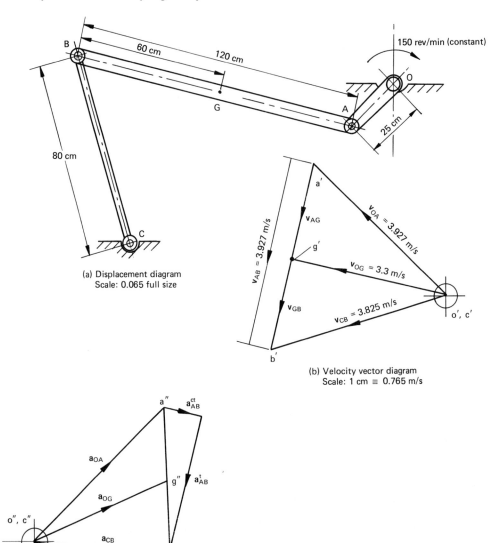

(a) Displacement diagram
Scale: 0.065 full size

(b) Velocity vector diagram
Scale: 1 cm ≡ 0.765 m/s

(c) Acceleration vector diagram
Scale: 1 cm ≡ 12 m/s²

Figure 3.7

to O and C are the same. Thus points o', c' and o'', c'' are the starting points of the velocity and acceleration vector diagrams respectively and are termed the *pole points*.

(a) *Velocity diagram* (see Figure 3.7b). Consider the velocity of A relative to O, i.e. v_{OA}. Since the length OA is constant, $\dot{r} = 0$ in equation 3.3. Thus $v_{OA} = v^t_{OA}$ and is represented by the vector of magnitude $(OA \times 150\pi/30) = 3.927$ m/s, i.e. $r\dot{\phi}$ in equation 3.3, and acting as shown.

Consider now $\mathbf{v}_{AB}$. Once again, since the length BA is constant, $\mathbf{v}_{AB} = \mathbf{v}_{AB}^t$. Although we know the line of action of $\mathbf{v}_{AB}^t$, i.e. perpendicular to AB, we do not at this stage know either its magnitude or direction. Consider, however, $\mathbf{v}_{CB} = \mathbf{v}_{CB}^t$ (since the length BC is constant). Once again, we know the line of action of $\mathbf{v}_{CB}^t$, perpendicular to BC, but neither its magnitude nor its direction. Therefore, if we draw the lines of action of $\mathbf{v}_{CB}^t$ and $\mathbf{v}_{AB}^t$, where they intersect will represent the terminal point of the velocity vector describing the point B, i.e. b'.

From the above, we can construct the velocity diagram for OABC, as shown in Figure 3.7b. To locate the terminal point of the vector describing the velocity of the point G on the diagram, i.e. g', we use the condition that A, G and B are all constant points on the link AGB, i.e.

$$\mathbf{v}_{AG} = \mathbf{v}_{AB} \times AG/AB$$

or

$$a'g' = a'b' \times AG/AB = 3.927 \times 60/120 = 1.9635 \, \text{m/s}^2$$

(b) *Acceleration diagram* (see Figure 3.7c). Once again starting at pole point (o″, c″), consider the acceleration of A relative to O. Since the crank OA is of constant length and rotates at a constant angular speed of 150 rev/min, then from equation 3.4, $\dot{r} = \ddot{r} = \ddot{\phi} = 0$. Therefore, in this case, we are left only with the centripetal component of the acceleration of magnitude $r\dot{\phi}^2$, which in the present situation is $OA \times (150\pi/30)^2 = 61.685 \, \text{m/s}^2$. The line of action of $\mathbf{a}_{OA}^{ct}$ is parallel to the line OA and acts in a direction away from A and towards O. On the acceleration vector diagram this is represented by the line o″a″.

Now consider the acceleration of B relative to A. Once again, since the length AB is constant, $\dot{r} = \ddot{r} = 0$ in equation 3.4, leaving us with only the centripetal and tangential components, i.e. $\mathbf{a}_{AB} = \mathbf{a}_{AB}^{ct} + \mathbf{a}_{AB}^t$. Now in this case

$$\dot{\phi} = \text{rotational speed of B about A} = |\mathbf{v}_{AB}^t|/AB$$

where $|\mathbf{v}_{AB}^t|$ is the magnitude of the vector $\mathbf{v}_{AB}^t = 3.927 \, \text{m/s}$ and $r = AB = 1.2 \, \text{m}$. Then

$$|\mathbf{a}_{AB}^{ct}| = |\mathbf{v}_{AB}^t|^2/AB = 3.927^2/1.2 = 12.85 \, \text{m/s}^2$$

acting along a line parallel to AB in a direction towards A. In the case of the tangential acceleration of B related to A, $\mathbf{a}_{AB}^t$, only its line of action (perpendicular to AB) can be drawn on the acceleration diagram at this stage.

Now consider $\mathbf{a}_{CB}$ and, since CB is of constant length, once again we are left only with the centripetal and tangential components, i.e. $\mathbf{a}_{CB} = \mathbf{a}_{CB}^{ct} + \mathbf{a}_{CB}^t$, such that

$$|\mathbf{a}_{CB}^{ct}| = |\mathbf{v}_{CB}^t|^2/CB = 3.825^2/0.8 = 18.288 \, \text{m/s}^2$$

and acts parallel to BC in a direction towards C. Once again, as in the case of $\mathbf{a}_{AB}^t$, we know only the line of action of $\mathbf{a}_{CB}^t$ (perpendicular to CB). However, if we now draw on the acceleration diagrams the line of action of $\mathbf{a}_{CB}^t$, then where it intersects with the line of action of $\mathbf{a}_{AB}^t$ corresponds to the point b″—the point representing the acceleration of B—and the line o″b″ represents the absolute acceleration vector of B, $\mathbf{a}_{OB}$ (or $\mathbf{a}_{CB}$).

Hence we have now completed the acceleration diagram for the mechanism OABC. As in the case of the velocity diagram, the point g″ can be obtained by

$$a''g'' = a''b'' \times AG/AB = 23.6 \, \text{m/s}^2$$

and the line drawn from o″ to g″ is the absolute acceleration vector of the point G, $\mathbf{a}_{OG} = 49.6 \, \text{m/s}^2$ in the direction shown.

The magnitude of the angular acceleration of link AGB, α_G, is $|a^t_{GB}|/GB$; and since A, G and B are on the rigid link AGB

$$\alpha_G = \frac{|a^t_{AB}|}{AB} = \frac{45.6}{1.2} = 38\,\text{rad/s}^2 \quad \text{(anticlockwise)}$$

Example 3.2

For the piston drive mechanism shown in Figure 3.8a, in which the crank OA rotates at a constant speed of 400 rev/min in the direction shown, construct the velocity and acceleration vector diagrams and hence obtain the absolute accelerations of piston D and centre of gravity, G, of link AC. Also find the angular acceleration of link AC.

SOLUTION

The reader should note that the sub-mechanism OACE is similar to that of OABC analysed in Example 3.1. Therefore, by referring to the solution of the previous example, the reader should be able to construct the velocity and acceleration vector diagrams for the sub-mechanism OACE which are contained within Figures 3.8b and 3.8c respectively. Similarly, as before, the reader should also be in a position to locate the points g' and b' on the velocity vector diagram, and g" and b" on the acceleration vector diagram, noting that A, G, B and C lie on the same solid link.

(a) *Velocity vector diagram* (see Figure 3.8b). Starting at the velocity point B', consider now the velocity of D relative to B, i.e. $v_{BD} = v^t_{BD}$ (since length BD is constant). At this stage we only know the line of action of v^t_{BD}, i.e. perpendicular to the line BD, which is drawn in on the velocity vector diagram passing through point B'.

Now, relative to the fixed space point, o',e', the absolute velocity of D, v_{OD}, must be a vector whose line of action is parallel to the cylinder wall at D, i.e. horizontal. Therefore, upon drawing in this horizontal line of action on the velocity vector diagram (passing through the point o',e'), where it intersects the line of action of $v_{BD}(=v^t_{BD})$ is the velocity point d', such that the line o'd' represents the absolute velocity of piston D, i.e. 5.35 m/s →.

(b) *Acceleration vector diagram* (see Figure 3.8c). Having established the point b" on the acceleration vector diagram of the sub-mechanism OACE, let us now consider a_{BD} such that $a_{BD} = a^{ct}_{BD} + a^t_{BD}$, since $\dot{r} = \ddot{r} = 0$ in equation 3.4 and

$$|a^{ct}_{BD}| = \frac{|v^t_{BD}|^2}{BD} = \frac{7.7^2}{0.61} = 97.197\,\text{m/s}^2$$

Once again, the line of action of a^{ct}_{BD} is parallel to BD and acts in a direction towards B as shown. Thus a^{ct}_{BD} can be immediately drawn in on the acceleration vector diagram. Having done so, we can now draw in the line of action a^t_{BD}, i.e. perpendicular to BD as shown.

Once again, as in the velocity vector diagram, the line of action of a_{OD} is horizontal passing through the point o",e". The intersection of the lines of action of a_{OD} and a^t_{BD} is the acceleration point d". Thus, from the completed acceleration vector diagram, $a_{OG} = 320\,\text{m/s}^2$ in the direction indicated, $a_{OD} = 164\,\text{m/s}^2 \rightarrow$, and $\alpha_G = |a^t_{AC}|/AC = 237.5/0.915 = 259.6\,\text{rad/s}^2$ (anticlockwise).

Example 3.3

Figure 3.9a shows a general coplanar mechanism in which the crank AB rotates at

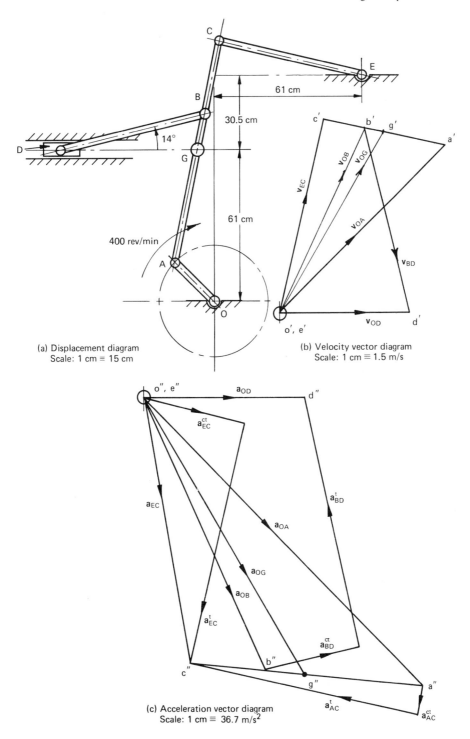

(a) Displacement diagram
Scale: 1 cm ≡ 15 cm

(b) Velocity vector diagram
Scale: 1 cm ≡ 1.5 m/s

(c) Acceleration vector diagram
Scale: 1 cm ≡ 36.7 m/s²

Figure 3.8

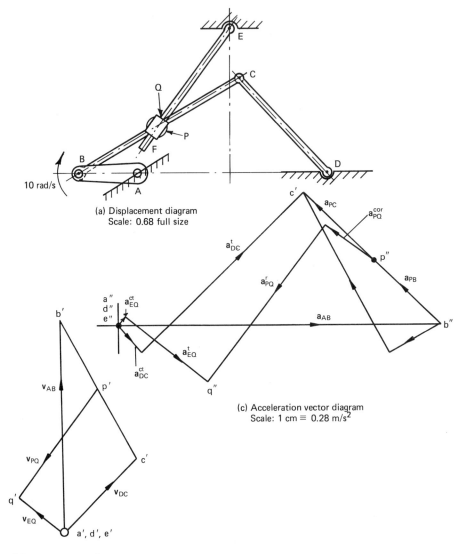

(a) Displacement diagram
Scale: 0.68 full size

(c) Acceleration vector diagram
Scale: 1 cm ≡ 0.28 m/s²

(b) Velocity vector diagram
Scale: 1 cm ≡ 0.044 m/s

Figure 3.9

a constant speed of 10 rad/s in the direction indicated. At the point P on the link BC is located a pivot which retains a swivel block through which the link EF is free to slide. For the instantaneous position shown, construct the velocity and acceleration vector diagrams for the complete mechanism.

SOLUTION

Let the point Q, indicated in Figure 3.9a, be a point on the link EF located, at the instant shown, *within* the swivel block supported at P.

Once again, by virtue of the experience gained in Example 3.1, the reader should be able to confirm the velocity and acceleration vector diagrams for the sub-mechanism ABCD as contained in Figures 3.9b and 3.9c respectively. Furthermore, since points B, P and C lie on the same solid link, the velocity and acceleration points, p' and p'' respectively, can be located.

(a) *Velocity vector diagram* (see Figure 3.9b). Starting at the point p', consider v_{PQ}. Since at the instant shown Q and P are coincidental, then in equation 3.3 $r = 0$. Thus, $v_{PQ} = v_{PQ}^r$ and the line of action of v_{PQ}^r (parallel to EQ) can be drawn on the velocity vector diagram, through the point p'. Consider now v_{EQ} such that $v_{EQ} = v_{EQ}^t$ (since EQ is of constant length). As in the case of v_{PQ}^r, we know only the line of action of v_{EQ}^t, i.e. perpendicular to the line EQ and passing through the point a', d', e' on the velocity vector diagram. Where the lines of action of v_{PQ}^r and v_{EQ}^t intersect is the velocity point q'.

(b) *Acceleration vector diagram* (see Figure 3.9c). Starting at the point p'', let us now consider a_{PQ} in accordance with equation 3.4. Since P and Q are coincident, $r = 0$; therefore $a_{PQ} = a_{PQ}^r + a_{PQ}^{Cor}$. Now the magnitude of the Coriolis component is

$$|a_{PQ}^{Cor}| = \frac{2 \times |v_{PQ}^r| \times |v_{EQ}^t|}{EQ} = 2\dot{r}\dot{\phi} \qquad \text{in equation 3.4}$$

$$= \frac{2 \times 0.162 \times 0.069}{0.05} = 0.447 \, \text{m/s}^2$$

and its direction is obtained by rotating the direction of the vector v_{PQ}^r through 90° in the direction of rotation of Q. By considering the direction of the vector v_{EQ}^t we deduce that Q is rotating in a clockwise sense, thus the direction of a_{PQ}^{Cor} is as shown in Figure 3.9c. Therefore, at the point p'' on the acceleration vector diagram we draw a_{PQ}^{Cor} together with the line of action of a_{PQ}^r (parallel to EF) as shown.

Consider now $a_{EQ} = a_{EQ}^{ct} + a_{EQ}^t$. Since $\dot{r} = \ddot{r} = 0$ in equation 3.4,

$$|a_{EQ}^{ct}| = \frac{|v_{EQ}^t|^2}{EQ} = \frac{0.069^2}{0.05} = 0.095 \, \text{m/s}^2$$

and acts in a direction parallel to the line EQ and towards E as shown in Figure 3.9c. Hence starting at the pole point a'', d'', e'' on the acceleration vector diagram we can draw a_{EQ}^{ct} and the line of action of a_{EQ}^t as shown. Subsequently, the point of intersection of the lines of action of a_{EQ}^t and a_{EQ}^t is the point q''.

3.2 Kinetics of general plane motion

So far, in Section 3.1, we have considered the kinematics of coplanar mechanisms, and in particular the acceleration of parts of the mechanisms. Obtaining the acceleration vectors is not the end product of the study, but rather a means to an end, since having determined these vectors at all parts of the mechanisms we can then proceed to calculate the force and moment vectors necessary to produce such acceleration vectors. Although the relationship between forces, moments and acceleration (linear and angular) is fully described by Newton's Second Law of Motion, for the present purposes it is more convenient to describe this law in a slightly different form. Consider the body shown in Figure 3.10 undergoing general plane motion by the action of applied force vectors, F_1 and F_2, such that the linear acceleration of the centre of mass of the body, G, is a_{OG} and the angular acceleration is α_G as shown.

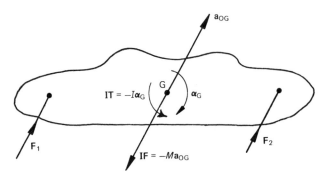

Figure 3.10

Denoting the mass and polar mass moment of inertia about G of the body as M and I respectively, from Newton's Second Law of Motion

$$\text{Net applied force} = \mathbf{F}_1 + \mathbf{F}_2 = \mathbf{F} = M\mathbf{a}_{OG} \tag{3.5}$$

and

$$\text{Moment of forces } \mathbf{F}_1 \text{ and } \mathbf{F}_2 \text{ about } G = \mathbf{T} = I\boldsymbol{\alpha}_G \tag{3.6}$$

Equation 3.5 can be rewritten as

$$\mathbf{F} - M\mathbf{a}_{OG} = 0 \tag{3.7}$$

The term $-M\mathbf{a}_{OG}$, contained in equation 3.7, is termed the *inertia force vector* (or *d'Alembert force vector*, after Jean le Rond d'Alembert (1717–1783), French philosopher and mathematician), **IF**, such that

$$|\mathbf{IF}| = M \times |\mathbf{a}_{OG}| \tag{3.8}$$

and whose line of action is that of $\mathbf{a}_{OG}$ but *acts in a direction directly opposite to that of* $\mathbf{a}_{OG}$ (see Figure 3.10). Equation 3.5 now becomes

$$\mathbf{F} + \mathbf{IF} = 0 \tag{3.9}$$

i.e. an equation that implies static equilibrium.

Similarly, consider equation 3.6 and rewrite it in the form

$$\mathbf{T} - I\boldsymbol{\alpha}_G = 0 \tag{3.10}$$

In this case the term $-I\boldsymbol{\alpha}_G$ is termed the *inertia torque vector*, **IT**, such that

$$|\mathbf{IT}| = I|\boldsymbol{\alpha}_G| \tag{3.11}$$

and acts in a sense opposite to that of $\boldsymbol{\alpha}_G$ (see Figure 3.10). Thus equation 3.6 can be rewritten as

$$\mathbf{T} + \mathbf{IT} = 0 \tag{3.12}$$

and is once again an equation implying static equilibrium.

The applied force and torque vectors, **F** and **T** respectively, can therefore be described as the force and torque vectors required to balance the inertia force and torque vectors acting on the body.

Example 3.4

For the general plane mechanism described in Example 3.2 find the necessary instantaneous torque on the shaft at O required to drive the mechanism and the magnitude of the forces acting on the bearings at O, A, B, C, D and E. Assume that links OA, BD and CE have negligible mass but that the piston at D, and link AC have the following parameters: $M_D = 18\,\text{kg}$, $M_{AC} = 23\,\text{kg}$ (G = centre of mass of link AC) and the link AC is of uniform cross-section. Neglect the effects of gravitational force and assume that all bearings are frictionless.

SOLUTION

Consider firstly the link DB. Since it has negligible mass and the bearings at B and D produce no frictional torques, then for equilibrium the forces acting on the link at B and D must be equal and opposite and act directly along the length of the link, i.e. $\mathbf{F_B} = -\mathbf{F_D}$ (see Figure 3.11).

In the absence of gravitational force, consider the forces acting on the piston D. From the acceleration vector diagram (Figure 3.8c) we see that the absolute acceleration of the piston is $164\,\text{m/s}^2 \rightarrow$. Therefore from equation 3.8, the inertia force vector, **IF**, is

$$\mathbf{IF} = 18 \times 164 = 2952\,\text{N} \leftarrow$$

Therefore the forces acting on the piston are as shown in Figure 3.11, where **R** is the normal reaction force vector between the piston and cylinder wall, and $\mathbf{F_D}$ is the

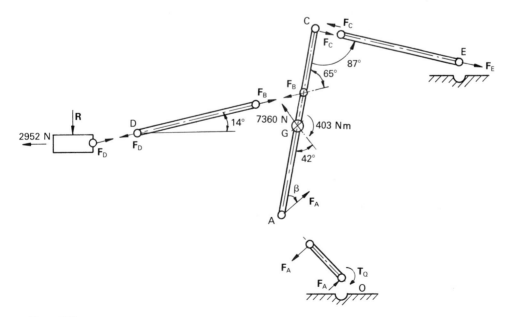

Figure 3.11

force vector exerted on the piston by the link BD. Thus

$$|\mathbf{F_D}|\cos 14° = 2952\,\text{N} \qquad \text{i.e.} \qquad |\mathbf{F_D}| = 3042.37\,\text{N}$$

and

$$|\mathbf{F_D}|\sin 14° = |\mathbf{R}| = 736\,\text{N}$$

Now consider the link AC with centre of mass G equidistant from A and C. From the acceleration vector diagram,

$$\mathbf{a_{OG}} = 320\,\text{m/s}^2 \searrow$$

Thus

$$\mathbf{IF_G} = 23 \times 320 = 7360\,\text{N} \nwarrow$$

and

$$\alpha_G = 259.6\,\text{rad/s}^2 \qquad \text{(anticlockwise)}$$

then

$$\begin{aligned}
\mathbf{IT} &= M_{AC} \times (AC^2/12) \times 259.6 \\
&= 23 \times (0.9^2/12) \times 259.6 \\
&= 403\,\text{N m} \qquad \text{(clockwise)}
\end{aligned}$$

Once again, since link CE is assumed massless and has frictionless bearings at C and E, the forces at C and E must act along the line CE. Therefore the forces and torques acting on link AC are described in Figure 3.11. Note that, although link OA is assumed massless, since there is the likelihood of a finite torque acting at O, the equal and opposite forces acting at points O and A will not act along the centreline of OA. Figure 3.11 shows the link AC in isolation with the forces and torque acting on it where β is the unknown angle describing the direction of $\mathbf{F_A}$ relative to the link AC. Resolving forces perpendicular to AC gives

$$|\mathbf{F_C}|\sin 87° + |\mathbf{F_A}|\sin \beta = 7360\sin 42° + 3042.37\sin 65°$$

thus

$$0.9986|\mathbf{F_C}| + |\mathbf{F_A}|\sin \beta = 7682\,\text{N} \qquad (1)$$

and resolving forces parallel to AC gives

$$|\mathbf{F_C}|\cos 87° + |\mathbf{F_A}|\cos \beta = 7360\cos 42° + 3042.37\cos 65°$$

thus,

$$0.0523|\mathbf{F_C}| + |\mathbf{F_A}|\cos \beta = 6755\,\text{N} \qquad (2)$$

Now taking moments about G gives

$$0.4575 \times |\mathbf{F_C}|\sin 87° + 403 = 0.4575 \times |\mathbf{F_A}|\sin \beta + 0.1525 \times 3042.37\sin 65°$$

thus

$$0.9987|\mathbf{F_C}| - |\mathbf{F_A}|\sin \beta = 38.12 \qquad (3)$$

Now from equations 1 and 3 we obtain

$$\begin{aligned}
|\mathbf{F_C}| &= \text{magnitude of the force acting on bearings at C and E} \\
&= 3865.28\,\text{N}
\end{aligned}$$

Hence from equations 1 and 2 we have

$\beta = \arctan(0.583) = 30.25°$

and

$|\mathbf{F_A}| = 3822.13/\sin 30.25°$
$= 7586\,\text{N}$
$= \text{force on the bearing at A}$

Referring now to link OA,

$\mathbf{T_Q} = \text{torque necessary to drive system}$
$= (7586 \sin 87°) \times \text{OA}$
$= 1742.4\,\text{N m} \qquad \text{(clockwise)}$

and the magnitude of the force acting on bearing at $0 = |\mathbf{F_A}| = 7586\,\text{N}$.

Problems

3.1   Figure 3.12 shows the needle drive mechanism of an industrial sewing machine. The needle is constrained by guides to move along the line XX. Links BC and CD are both clamped to shaft C, so that these links move together as one.

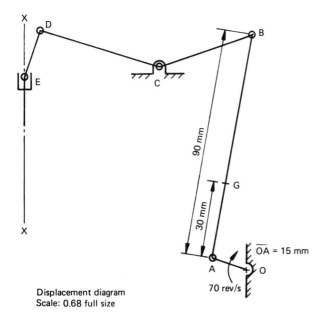

Displacement diagram
Scale: 0.68 full size

Figure 3.12

Draw a velocity diagram for the mechanism at the instant shown in the figure. Hence determine the needle velocity at this instant, and state whether the needle is rising or falling.

The centre of gravity of link AB is at G. The mass of link AB is 20 g and the moment of inertia of link AB about G is 0.04 kg m². Find also the total kinetic energy of AB at the instant shown.

Answer: 10 m/s downwards; 65.6 J.

3.2 Figure 3.13 shows a simple sliding mechanism in which the piston at B is driven via a connecting rod AB, by a crank OA which rotates at a constant angular velocity Ω rad/s in an anticlockwise direction.

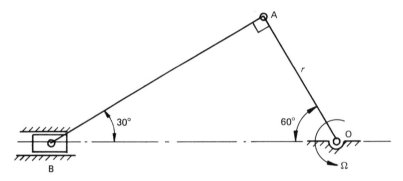

Figure 3.13

For the position indicated, where the crank OA is perpendicular to the connecting rod AB, construct the velocity and acceleration vector diagrams for the mechanism, expressing the vectors in terms of the parameters r and Ω, and determine:

(a) the acceleration of the piston,
(b) the force on the cylinder wall at B,
(c) the torque applied to the crack at O to overcome the inertia effects associated with the piston which has a mass M (the mass of the crank OA and connecting rod AB may be considered negligible).

Show also that the power transmitted by the mechanism is proportional to Ω^3.

Answer: $\frac{2}{9}\Omega^2 r \rightarrow$; $\frac{2M}{9\sqrt{3}}\Omega^2 r\uparrow$; $\frac{4M\sqrt{3}}{27}\Omega^2 r\searrow$.

3.3 Figure 3.14 shows a schematic diagram of a mechanism in which crank OA rotates at 1200 rev/min in a clockwise direction. Determine the acceleration of the sliding block in magnitude and direction.

Answer: 233 m/s² at 16.8° anticlockwise from horizontal.

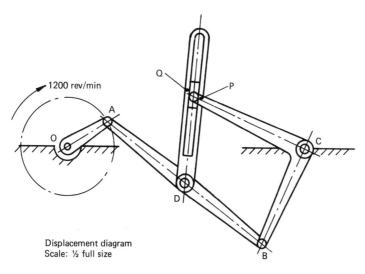

1200 rev/min

Displacement diagram
Scale: ½ full size

Figure 3.14

4

Turning moment diagrams and flywheel design

4.1 Turning moment diagrams

Consider the single-cylinder engine system shown in Figure 4.1 where the pressure, p, acting on the piston varies according to the cycle stage, i.e. compression, power, exhaust and suction. Furthermore, this pressure, or more specifically pressure force, performs three specific functions, namely:

(1) to provide the necessary forces and torques to accelerate the various components of the system, i.e. piston, connecting rod and crank;
(2) to provide the necessary force to overcome the friction within the system; and
(3) to provide the net driving torque, T_Q, at the crankshaft, O.

The torque, T_Q is therefore the torque, or turning moment, available to drive some load or machine. By virtue of the cyclic nature of the pressure force, and from Section 3.2, the cyclic nature of the inertia forces, torques and friction forces in such mechanisms, the reader will appreciate that the net driving torque will also be cyclic in nature, i.e. fluctuate over one cycle of operation, therefore $T_Q = T_Q(\theta)$, where θ is the crank angle with respect to some fixed reference. Figures 4.2a and 4.2b show typical forms of $T_Q(\theta)$ over one cycle of operation, for a single-cylinder four-stroke and four-cylinder four-stroke engine respectively.

The reader should note that, for the single-cylinder arrangement, the cycle angle $\theta_Q = 4\pi$, and for the four-cylinder configuration $\theta_Q = \pi$.

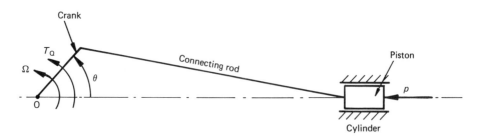

Figure 4.1

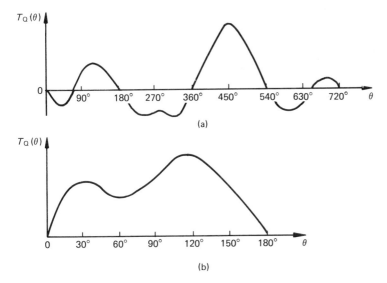

Figure 4.2

4.1.1 Complete system cycle angle

Let us consider the case of an engine whose $T_Q(\theta)$ versus θ characteristics are known, which is required to drive some machine whose load torque, T_L, is defined and is also cyclic in nature, i.e. $T_L = T_L(\theta)$, but its cycle angle, θ_L, is different from θ_Q. Therefore, the first parameter we are required to establish is the cycle angle for the completed system, θ_S. We start by stating that every cycle of the complete system is identical and repeats itself. Therefore we can write

$$\theta_S = m\theta_Q = n\theta_L \tag{4.1}$$

where m and n are the *lowest integer values* that would satisfy equation 4.1. For example,

(1) $\theta_Q = 2\pi$, $\theta_L = 4\pi$, therefore

   $m2\pi = n4\pi$

   thus $m = 2$, $n = 1$, giving $\theta_S = 4\pi$.
(2) $\theta_Q = \pi$, $\theta_L = 3\pi/2$, therefore

   $m\pi = n3\pi/2$

   thus $m = 3$, $n = 2$, giving $\theta_S = 3\pi$.

4.1.2 Complete system turning moment diagram

Let us now consider the hypothetical case where a four-cylinder four-stroke engine having $T_Q(\theta)$ versus θ characteristics similar to that shown in Figure 4.2b drives a machine whose $T_L(\theta)$ versus θ characteristics over its cycle angle $\theta_L = 2\pi$, as shown in Figure 4.3. Hence superimposing Figure 4.2b and Figure 4.3 we obtain the complete system turning moment diagram as shown in Figure 4.4. Note that, since $\theta_Q = \pi$ and $\theta_L = 2\pi$, $\theta_S = m\pi = n2\pi$, i.e. $n = 1$, $m = 2$; hence $\theta_S = 2\pi$.

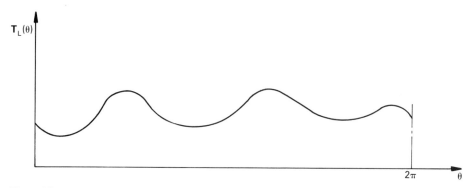

Figure 4.3

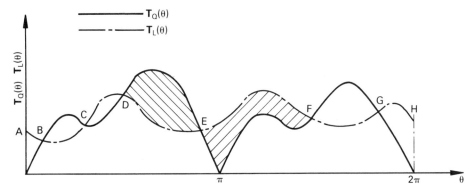

Figure 4.4

Now applying Newton's Second Law over the cycle angle, θ_S, we have

$$T_Q(\theta) - T_L(\theta) = I\frac{d\Omega}{dt} = I\Omega\frac{d\Omega}{d\theta} \tag{4.2}$$

where I is the total combined moment of inertia of the complete system and Ω is the rotational speed of the system at some value of θ between 0 and θ_S. Hence

$$\int_0^{\theta_S} [T_Q(\theta) - T_L(\theta)]\,d\theta = I\int_{\Omega_A}^{\Omega_H} \Omega\,d\Omega = \tfrac{1}{2}I(\Omega_H^2 - \Omega_A^2) \tag{4.3}$$

where Ω_A and Ω_H are the system speeds at the beginning ($\theta = 0$) and end ($\theta = \theta_S$) of the system cycle respectively. Now if the left-hand side of equation 4.3 is not equal to 0, this would imply that $\Omega_H \neq \Omega_A$, i.e. the system would be gaining (or losing) kinetic energy from cycle to cycle. In the present text, however, we will only consider the case where $\Omega_A = \Omega_H$, i.e. the system speeds at the beginning and end of the cycle are the same. Therefore, for such a condition, equation 4.3 implies that

$$\int_0^{\theta_S} T_Q(\theta)\,d\theta = \int_0^{\theta_S} T_L(\theta)\,d\theta \tag{4.4}$$

or the area under the $T_Q(\theta)$ versus θ plot (work done by the engine over θ_S) must equal the area under the $T_L(\theta)$ versus θ plot (work done on the machine over θ_S). A note of interest is that, if $T_L(\theta) = T_L$, a constant, then, for $\Omega_H = \Omega_A$,

$$T_L = \frac{1}{\theta_S} \int_0^{\theta_S} T_Q(\theta)\, d\theta = \text{mean value of } T_Q(\theta)$$

Let us now consider some intermediate part of the cycle, say between points D and E as shown in Figure 4.4. Between these two intersect points we observe that $T_Q(\theta) \geqslant T_L(\theta)$, so once again applying Newton's Second Law between these two points

$$\int_{\theta_D}^{\theta_E} [T_Q(\theta) - T_L(\theta)]\, d\theta = \tfrac{1}{2}I(\Omega_E^2 - \Omega_D^2) \qquad (4.5)$$

and since

$$\int_{\theta_D}^{\theta_E} T_Q(\theta)\, d\theta > \int_{\theta_D}^{\theta_E} T_L(\theta)\, d\theta$$

then

$$\tfrac{1}{2}I(\Omega_E^2 - \Omega_D^2) = \text{gain in kinetic energy of the system between cycle points D and E}$$
$$= \text{cross-hatched area in Figure 4.4}$$

Likewise, between E and F where $T_Q(\theta) \leqslant T_L(\theta)$,

$$\int_{\theta_E}^{\theta_F} [T_Q(\theta) - T_L(\theta)]\, d\theta = \tfrac{1}{2}I(\Omega_E^2 - \Omega_F^2) \qquad (4.6)$$

$$= \text{loss in kinetic energy of the system between cycle}$$
$$\text{points E and F}$$
$$= \text{cross-hatched area in Figure 4.4}$$

Thus, if we know the values of all such areas of the system turning moment diagram, we can quantify the fluctuation in system kinetic energy over the complete system cycle.

For example, denoting the system kinetic energy at the beginning of the cycle (point A) as KE_A, then (since $T_Q(\theta) \leqslant T_L(\theta)$)

Kinetic energy at $B = KE_B = KE_A - $ enclosed area between A and B
$$= KE_A - \text{Area}_{AB}$$

Similarly,

$$KE_C = KE_A - \text{Area}_{AB} + \text{Area}_{BC}$$
$$KE_D = KE_A - \text{Area}_{AB} + \text{Area}_{BC} - \text{Area}_{CD}$$
$$\vdots$$
$$KE_H = KE_A - \text{Area}_{AB} + \text{Area}_{BC} - \text{Area}_{CD} + \text{Area}_{DE} - \text{Area}_{EF} + \text{Area}_{FG} - \text{Area}_{GH}$$
$$= KE_A \quad (\text{since } \Omega_A = \Omega_H)$$

The maximum fluctuation in kinetic energy (MFKE) of the system over the cycle will be the difference in kinetic energy associated with the points which represent the highest and the lowest values of kinetic energy.

Example 4.1

A multi-cylinder engine produces a net driving torque, $T_Q(\theta)$, given by the equation

$$T_Q(\theta) = (500 + 20 \sin 3\theta)\, \text{N m}$$

where θ is a crank angle relative to some fixed reference. It is required to drive a machine that represents a load torque $T_L(\theta)$ of the form

$$T_L(\theta) = (A + 60 \cos 2\theta) \, \mathrm{N} \, \mathrm{m}$$

where A is a variable. Calculate:

(a) the cycle angle, θ_S, for the complete system,
(b) the value of A necessary to ensure that the system speeds at the beginning and at the end of the cycle are the same, and of value 20 rad/s,
(c) the maximum fluctuation in kinetic energy during the cycle,
(d) the maximum percentage fluctuation in speed over the cycle, given that the system moment of inertia is $30 \, \mathrm{kg} \, \mathrm{m}^2$,
(e) an estimate of the power required to drive the system.

SOLUTION

(a) Figures 4.5a and 4.5b show plots of $T_Q(\theta)$ and $T_L(\theta)$ versus θ over one cycle of each respectively; hence $\theta_Q = 2\pi/3$, $\theta_L = \pi$, and (from equation 4.1)

$$m \, 2\pi/3 = n\pi = \theta_S$$

i.e. $m = 3$, $n = 2$, giving $\theta_S = 2\pi$.

(b) If system speed at the beginning and at the end of cycle ($\theta = 2\pi$) are the same, then from equation 4.4

$$\int_0^{2\pi} (500 + 20 \sin 3\theta) \, \mathrm{d}\theta = \int_0^{2\pi} (A + 60 \cos 2\theta) \, \mathrm{d}\theta$$

giving

$$A = 500 \, \mathrm{N} \, \mathrm{m}$$

(c) Figure 4.5c shows the complete system turning moment diagram over one cycle, $\theta_S = 2\pi$. For intersect points B, C, D and E, $T_Q(\theta) = T_L(\theta)$, giving $\sin 3\theta = 3 \cos 2\theta$, and iterating gives $\theta_B = 0.623$, $\theta_C = 2.518$, $\theta_D = 4.012$ and $\theta_E = 5.413$ rad. Therefore

$$\mathrm{Area}_{AB} = \int_0^{0.623} [T_Q(\theta) - T_L(\theta)] \, \mathrm{d}\theta$$

$$= \int_0^{0.623} (20 \sin 3\theta - 60 \cos 2\theta) \, \mathrm{d}\theta = -19.806 \, \mathrm{N} \, \mathrm{m}$$

Note the minus sign, since $T_Q(\theta) \leqslant T_L(\theta)$ over the range $0 < \theta < 0.623$.
Similarly,

$$\mathrm{Area}_{BC} = +52.946 [T_Q(\theta) \geqslant T_L(\theta)] \qquad 0.623 < \theta < 2.518$$
$$\mathrm{Area}_{CD} = -61.791 [T_Q(\theta) \leqslant T_L(\theta)] \qquad 2.518 < \theta < 4.012$$
$$\mathrm{Area}_{DE} = +70.637 [T_Q(\theta) \geqslant T_L(\theta)] \qquad 4.012 < \theta < 5.413$$
$$\mathrm{Area}_{EF} = -41.986 [T_Q(\theta) \leqslant T_L(\theta)] \qquad 5.413 < \theta < 2\pi$$

noting that $\sum \mathrm{Area} = 0$, which is in accordance with the condition that $\Omega_A = \Omega_F$. Now let the system kinetic energy at $A = KE_A$, hence

$$KE_B = KE_A - 19.806 \, \mathrm{N} \, \mathrm{m}$$
$$KE_C = KE_A - 19.806 + 52.946 = KE_A + 33.14 \, \mathrm{N} \, \mathrm{m}$$
$$KE_D = KE_A + 33.14 - 61.791 = KE_A - 28.651 \, \mathrm{N} \, \mathrm{m} \tag{1}$$

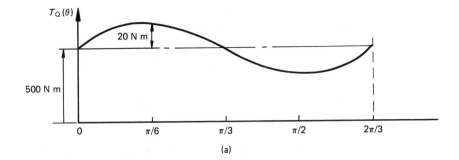

(a)

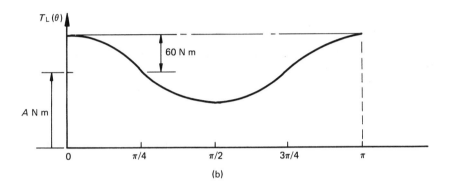

(b)

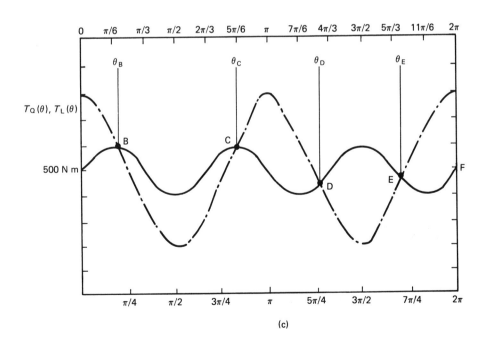

(c)

Figure 4.5

$$KE_E = KE_A - 28.651 + 70.637 = KE_A + 41.986 \, \text{N m} \tag{2}$$

$$KE_F = KE_A + 41.986 - 41.986 = KE_A \quad \text{(as expected)}$$

From the above, we deduce that the maximum value of kinetic energy is $KE_E = KE_A + 41.986 \, \text{N m}$, and the minimum value is $KE_D = KE_A - 28.651 \, \text{N m}$. Therefore

Maximum fluctuation of kinetic energy, $\text{MFKE} = KE_E - KE_D$

$$= 70.637 \, \text{N m}$$

(d) $\text{MFKE} = 70.637 \, \text{N m} = \frac{1}{2}I(\Omega_E^2 - \Omega_D^2)$, where Ω_E and Ω_D are the maximum and minimum system speeds respectively over the cycle. Therefore we can write the above equation as

$$\text{MFKE} = \frac{1}{2}I(\Omega_E + \Omega_D)(\Omega_E - \Omega_D) = I\Omega_m\Delta$$

where Ω_m is the mean value of Ω_E and Ω_D, and $\Delta \ (= \Omega_E - \Omega_D)$ is the maximum fluctuation of system speed over the cycle.
Alternatively we can write

$$\left(\frac{\text{MFKE}}{\Omega_m^2 I}\right) \times 100\% = \left(\frac{\Delta}{\Omega_m}\right) \times 100\% \tag{3}$$

$$= \text{maximum percentage fluctuation of speed (MPFS) about the mean speed } \Omega_m$$

Now, from equation 2,

$$I\Omega_E^2 = \frac{1}{2}I\Omega_A^2 + 41.986$$

where $\Omega_A = 20 \, \text{rad/s}$ and $I = 30 \, \text{kg m}^2$, i.e.

$$\Omega_E = \sqrt{\left(400 + \frac{41.986 \times 2}{30}\right)} = 20.06986 \, \text{rad/s}$$

Similarly, from equation 1,

$$\Omega_D = \sqrt{\left(400 - \frac{28.651 \times 2}{30}\right)} = 19.9522 \, \text{rad/s}$$

Therefore

$$\Omega_m = \frac{20.06986 + 19.9522}{2} = 20.011 \, \text{rad/s}$$

and, from equation 3,

$$\text{MPFS} = \frac{70.637 \times 100}{20.011^2 \times 30} = 0.588\%$$

$$= \pm 0.294\% \text{ about mean speed}$$

(e)

Power required = mean engine torque × mean speed

where

$$\text{Mean engine torque} = \frac{1}{2\pi} \int_0^{2\pi} (500 + 20 \sin 3\theta) \, d\theta$$
$$= 500 \, \text{N m}$$

and we will assume in this case that the mean speed is the speed at the beginning and end of cycle, i.e. 20 rad/s. Hence

Power required $= 500 \times 20 = 10 \, \text{kW}$

4.2 Flywheel design

A flywheel is a device, normally a circular member, which when added to an engine/machine system increases the total moment of inertia of the system. Consider once again equation 3, i.e.

$$\text{MPFS} = \frac{\text{MFKE}}{\Omega_m^2 I} \times 100\%$$

Now both MFKE and Ω_m, as we have already seen, are solely functions of the complete system turning moment diagram. Therefore the addition of a flywheel has the effect of reducing the MPFS to a permissible level. In this case, therefore,

$I = $ moment of inertia of system (I_S) + moment of inertia of flywheel (I_f)

Example 4.2

If, in Example 4.1, it was required to reduce the MPFS to 0.2 per cent (from 0.588 per cent), by the addition of a flywheel, what would be the required polar mass moment of inertia of the flywheel?

SOLUTION

From equation 3 in Example 4.1 we have

$$0.2 = \frac{70.637 \times 100}{20.011^2 \times I}$$

giving $I = 88.2 \, \text{kg m}^2$. Therefore,

$I_f = 88.2 - I_S = 88.2 - 30 = 58.2 \, \text{kg m}^2$

As stated previously, flywheels are normally circular members, and the designer must ensure that the stresses present in them due to the centripetal acceleration do not give rise to a situation of yielding. Generally flywheels are of solid circular or annular form.

4.2.1 Solid circular flywheel

Consider the solid circular flywheel of outer radius b, shown in Figure 4.6a, rotating at Ω rad/s about its central axis OO.

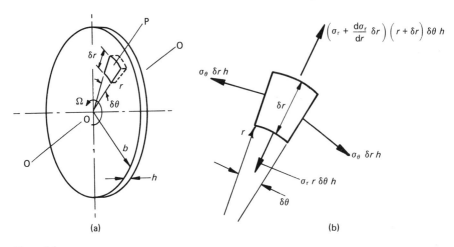

Figure 4.6

Now consider the small particle, P, of mass $\rho h r \, \delta r \, \delta\theta$, where ρ is the mass density of the flywheel material. The acceleration of this small mass will be $\Omega^2 r$ acting *towards* O along the line OP. The forces acting on this particle producing this acceleration are those shown in Figure 4.6b, where σ_r and σ_θ are the radial and hoop stresses respectively. Therefore, applying Newton's Second Law to this particle, we have (neglecting terms containing $\delta r \, \delta r \, \delta\theta$)

$$-\frac{d\sigma_r}{dr} r \, \delta r \, \delta\theta \, h - \sigma_r \, \delta r \, \delta\theta \, h + 2\sigma_\theta \, \delta r \, h \frac{\delta\theta}{2} = \rho \, \delta r \, \delta\theta \, h \Omega^2 r^2$$

giving

$$\sigma_\theta - \sigma_r - r\frac{d\sigma_r}{dr} = \rho \Omega^2 r^2 \qquad (4.7)$$

The reader will be aware that the presence of such stresses give rise to complementary strains ε_r and ε_θ respectively, and, if u is the radial displacement (or shift) of the point P due to σ_r and σ_θ, then

$$\varepsilon_\theta = \frac{u}{r} \qquad \text{and} \qquad \varepsilon_r = \frac{du}{dr} \qquad (4.8)$$

and from Hooke's Law

$$E\varepsilon_r = E\frac{du}{dr} = \sigma_r - v\sigma_\theta \qquad (4.9a)$$

and

$$E\varepsilon_\theta = E\frac{u}{r} = \sigma_\theta - v\sigma_r \qquad (4.9b)$$

where E is Young's modulus of elasticity, and v is Poisson's ratio. Hence obtaining du/dr from equation 4.9b and equating to equation 4.9a gives

$$(\sigma_\theta - \sigma_r)(1 + v) + r\frac{d\sigma_\theta}{dr} - vr\frac{d\sigma_r}{dr} = 0 \tag{4.10}$$

Combining equations 4.7 and 4.10 and solving the resultant differential equation, we obtained general solutions for σ_r and σ_θ in the form

$$\left.\begin{aligned}
\sigma_r &= A - \frac{B}{r^2} - \frac{(3+v)}{8}\rho\Omega^2 r^2 \\[2mm]
\sigma_\theta &= A + \frac{B}{r^2} - \frac{(1+3v)}{8}\rho\Omega^2 r^2
\end{aligned}\right\} \tag{4.11}$$

where A and B are arbitrary constants of integration. For a solid disc, since σ_r (and σ_θ) $\not\to \alpha$ as $r \to 0$, then $B = 0$. Also since, at $r = b$, $\sigma_r = 0$, we have

$$\sigma_r = \frac{\rho\Omega^2}{8}(3+v)(b^2 - r^2) = \frac{\rho\Omega^2}{8}(3+v)b^2\left(1 - \frac{r^2}{b^2}\right)$$

$$\sigma_\theta = \frac{\rho\Omega^2}{8}[(3+v)b^2 - (1+3v)r^2] \tag{4.12}$$

$$= \frac{\rho\Omega^2}{8}(3+v)b^2\left[1 - \frac{1+3v}{3+v}\cdot\frac{r^2}{b^2}\right]$$

Now, from the Tresca yield criterion, for this particular system,

$$\sigma_r \geqslant \sigma_r(\text{or } \sigma_\theta) \text{ at } r/b = 0 \tag{4.13}$$

where σ_y is the yield stress of the material, and from equation 4.12

$$\sigma_y \geqslant \frac{\rho\Omega^2(3+v)b^2}{8} \tag{4.14}$$

Equation 4.14 describes one design criterion for such a flywheel; the other criterion is that

$$I_f = \text{polar mass moment of inertia}$$
$$= \rho\pi h b^4/2 \tag{4.15}$$

Example 4.3

A solid circular steel flywheel is required to have a polar mass moment of inertia of 50 kg m^2 and is to be designed for a maximum operation speed of 400 rad/s. Determine suitable dimensions for the flywheel for a factor of safety of 2 given that, for steel, $\rho = 7600$ kg/m, $v = 0.3$ and yield stress $\sigma_y = 270$ MN/m^2.

SOLUTION

Since the required factor of safety is 2, we will design the flywheel for a yield stress of $\sigma_y/2$, i.e. 135 MN/m^2. From equation 4.14,

$$135 \times 10^6 \geqslant 7600 \times 400^2 \times b^2 \times 3.3/8 \Rightarrow b \leqslant 0.518 \text{ m}$$

and from equation 4.15

$$50 = 7600\pi h \times 0.518^4/2 \quad \Rightarrow \quad h = \text{flywheel thickness} = 5.8\,\text{mm}$$

In practice, the designer would reduce b and increase h accordingly.

4.2.2 Annular (or ring) flywheel

Figure 4.7a shows a typical annular flywheel carried by four spokes. For the present it will be assumed that the spokes carry zero load and the ratio t/b is small, therefore radial stress σ_r is negligible.

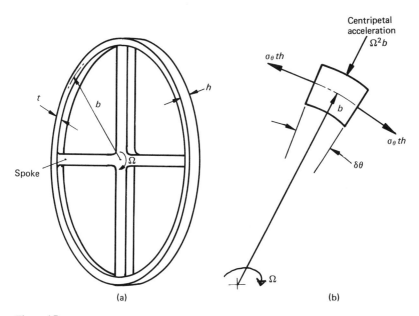

Figure 4.7

Consider a small element of the annulus shown in Figure 4.7b. Once again, applying Newton's Second Law,

$$2\sigma_\theta th\,\delta\theta/2 = \rho htb\,\delta\theta\Omega^2 b$$

or

$$\sigma_\theta = \rho\Omega^2 b \tag{4.16}$$

and from the Tresca yield criterion

$$\sigma_y \geqslant \rho\Omega^2 b^2 \tag{4.17}$$

The other design criterion is that

$$I_f = 2\pi\rho htb^3 \tag{4.18}$$

Example 4.4

Select suitable sizes for an annular steel flywheel which could replace the solid circular flywheel of Example 4.3.

SOLUTION

Once again, for a factor of safety of 2, from equation 4.17,

$$135 \times 10^6 \geqslant 7600 \times 400^2 \times b^2 \quad \Rightarrow \quad b < 0.33\,\text{m}$$

and from equation 4.18

$$50 = 2\pi \times 7600 \times 0.33^3 th \quad \Rightarrow \quad th = 0.02914\,\text{m}^2$$

If we set $t = h$,

$$t = h = 170.69\,\text{mm}$$

4.2.3 Additional design considerations

In Sections 4.2.1 and 4.2.2 flywheel dimensions were deduced solely on the basis of the flywheel having a certain polar mass moment of inertia, I_f, and the stresses within it do not give rise to the limit for yield being exceeded. In practice there are additional considerations such as (a) fatigue, which may arise due to speed fluctuations and vibration, and (b) shear stresses which manifest themselves at the interface between the flywheel and the shaft supporting it; such stresses can be severe at the point of 'start-up' where large inertia torques can be produced. Also, in the case of a motor car, large shear stresses in the same area can result when the clutch plate is engaged onto the flywheel. Normally a flywheel is 'keyed' to the supporting shaft as shown in Figure 4.8.

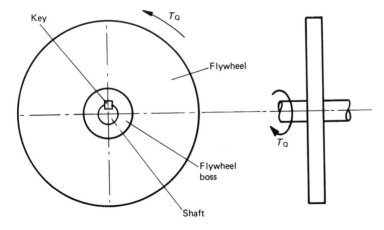

Figure 4.8

If the radius of the supporting shaft is r, then for a torque, T_Q,

$$\text{Force on key} = \frac{T_Q}{r}$$

$$\leqslant \text{shear area of key} \times \text{maximum shear stress of key material}$$

The torque, T_Q, could be inertia torque and/or clutch 'take-up' torque.

Problems

4.1   Figure 4.9 shows a torsiograph record taken from a diesel-driven electric generator set. The scales of the graph are:

Torque: 1 mm $\equiv$ 50 N m
Crank angle: 1 mm $\equiv$ 3°

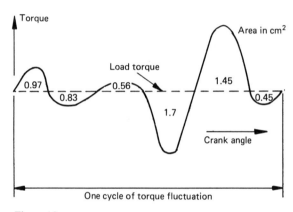

Figure 4.9

The moment of inertia of the rotating parts is 0.65 kg m^2. The speed fluctuation is required to be limited to ± 0.25 per cent of the mean speed which is 3000 rev/min.
Determine the moment of inertia of the additional flywheel required.

Answer: 0.395 kg m^2.

4.2   A diesel engine drives an electric generator which imposes a constant torque. A torque versus crank angle diagram for the engine constructed from torsiograph measurements indicates that the maximum fluctuation in energy during a cycle of torque variation is represented by an area of 1.75 cm^2 on the diagram. The diagram scales are as follows:

Torque: 1 cm $\equiv$ 500 N m
Crank angle: 1 cm $\equiv$ 30°

The mean speed is 1200 rev/min and the moment of inertia of the rotating parts is 0.73 kg m². Calculate the maximum percentage fluctuation of speed.

Answer: ± 1.987%.

4.3   A four-stroke single-cylinder internal combustion engine drives an electric generator at a steady mean speed of 3500 rev/min. The generator, which imposes a constant load torque on the engine, has an output of 10 kW and an efficiency of 86 per cent.

The engine torque may be taken as being zero at all times other than during the working stroke. During the working stroke the shape of the torque versus crank angle diagram is approximately that of a rectangle with a semicircle on top, as shown in Figure 4.10. Determine:

(a)   the peak engine torque,
(b)   the maximum fluctuation of kinetic energy,
(c)   the moment of inertia of the rotating parts to limit the speed fluctuation to ± 3.6 per cent of the mean speed.

Answer: 127.23 N m; 300 N m; 0.031 kg m².

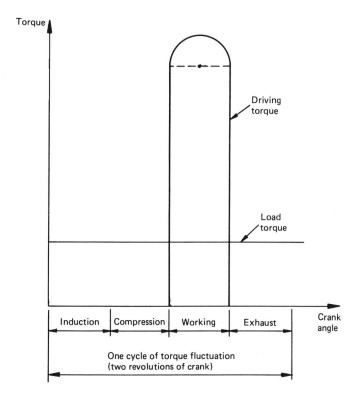

Figure 4.10

5

Applications of friction

5.1 Introduction

In the course of the design of almost all forms of machinery, the designer at some stage has to consider the effect of friction. Friction can be defined as basically that force which arises when two bodies in contact with each other move, or try to move, relative to each other whilst still in contact. The situation which inevitably occurs is often twofold, namely:

(1) force(s) are produced which oppose the relative motion (or intended motion), and
(2) heat is generated due to the sliding contact of the two bodies when relative motion is achieved.

To prehistoric man, the effect of (2) was a revelation to say the least, as he soon discovered that by rubbing two pieces of wood or flint vigorously together he could produce sufficient heat to initiate combustion, hence fire. However, as far as the present day design engineer is concerned, the heat generated by friction must very often be avoided at all costs, since, if it is allowed to occur unabated it may well give rise to machine 'seizure'. Let us consider, however, the forces produced as described in (1) above.

Consider a flat board which we will assume to be massless, of some material, lying on a flat horizontal table of either the same or some other material. Let us now spread a mass, M, of sand over the top of the board as shown in Figure 5.1. Thus, the normal reaction, N_r, between the underside of the board and the table will be Mg as shown, where g is gravitation acceleration.

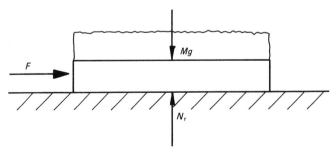

Figure 5.1

Let us now apply a force, F, as shown and note the value of this force which (a) just causes the load to slide (F_1), i.e. initiates motion, and (b) having initiated motion maintains motion at some *constant speed* (F_2). If this simple experiment is carried out, it will be observed that the force F_1 is greater than the force F_2. If we now continue the experiment by gradually increasing the weight of sand (Mg) and plot the corresponding value of F_1 and F_2 to a base of $N_r(= Mg)$. We would find that both these forces would bear a linear relationship to the normal reaction as shown in Figure 5.2.

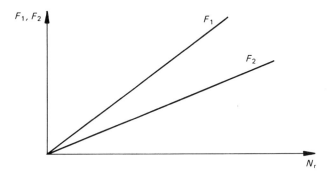

Figure 5.2

From the results of this simple experiment, we deduce that

$$F_1 \propto N_r \quad \text{i.e.} \quad F_1 = \mu_s N_r \tag{5.1a}$$

and

$$F_2 \propto N_r \quad \text{i.e.} \quad F_2 = \mu_k N_r \tag{5.1b}$$

where μ_s and μ_k are the static and kinetic coefficients of friction respectively, and $\mu_s > \mu_k$. These coefficients, as the reader will appreciate, are primarily functions of the contacting surface condition, i.e. roughness and cleanliness, and the types of contacting materials. Furthermore, experiments can be conducted on the simple system described above, which show that within reasonable bounds:

(1)   The forces $F_1 (= \mu_s N_r)$ and $F_2 (= \mu_k N_r)$ are independent of the area of the contact surface for a given normal reaction, N_r. (This becomes invalid if the area is so small that the pressure on the surface causes changes within the contacting surface condition, thus changing μ_s and μ_k.)
(2)   The force F_2 is independent of the velocity of slipping. (This becomes invalid as the velocity approaches zero.)

Prior to investigating specific considerations and applications of friction let us examine more closely the nature of the forces F_1 and F_2 described above.

5.1.1   Forces preventing slip

Consider a body held in contact with a surface by the normal free body forces of magnitude N_r which are normal to the contacting surfaces as shown in Figure 5.3.

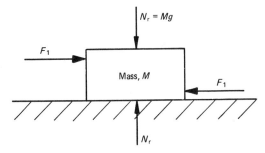

Figure 5.3

If we now apply some force of magnitude F_1 to the body as shown, then if slip does not occur, by Newton's First Law, this force must be opposed by a force due to the frictional resistance force of magnitude F_1, acting at the interfacing surface in a direction opposite to the direction of the applied force. This situation will be maintained until the value of F_1 has attained a value of $\mu_s N_r$, at which point slip will occur. Consequently, for $0 < F_1 < \mu_s N_r$, the friction resistance force must be equal in magnitude but opposite in direction to the applied force. Although the reader may consider the above statements to be obvious, it is well worth re-asserting them since, for example, when we consider belt drive systems in Section 5.2, the requirement for design against slip is based on the above.

5.1.2 Friction forces acting on moving bodies

Consider two bodies (A and B) held in contact under the action of the normal free body forces of magnitude N_r as shown in Figure 5.4. Additionally let us consider the case where the bodies are travelling at velocities of magnitude v_a and v_b. For the present let us assume that $v_a > v_b$ and that motion is sustained by the action of the forces of magnitude F_a and F_b as shown.

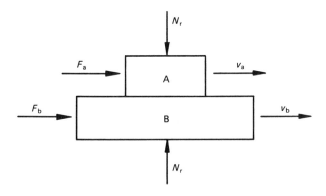

Figure 5.4

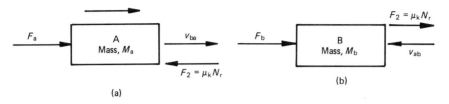

Figure 5.5

Now since $v_a \neq v_b$ slip must be occurring, and in consequence the magnitude of the force, F_2, acting along the interface between the two bodies, is

$$F_2 = \mu_k N_r$$

Before applying Newton's Second Law of Motion to each of the bodies let us confirm the direction in which F_2 acts on each of the bodies. This may be deduced from the following:

The friction force F_2, applied to a body, acts in a direction which opposes that of the velocity vector of the body relative to that of the body with which it is in contact.

Using the above statement, let us now write down the form of Newton's Second Law as appropriate to the body A. Figure 5.5a shows the body A, of mass M_a, subjected to the force of magnitude F_a, and the velocity vector of A relative to B, i.e. $v_{ba} = v_a - v_b$ in the direction shown, remembering that $v_a > v_b$. Thus, by the above statement, the friction force of magnitude $\mu_k N_r$ must act in the direction indicated, hence

$$F_a - \mu_k N_r = M_a \frac{dv_a}{dt} \tag{5.2a}$$

Similarly for B, as shown in Figure 5.5b,

$$F_b + \mu_k N_r = M \frac{dv_b}{dt} \tag{5.2b}$$

Let us now turn our attention to some selected examples involving the applications of friction.

5.2 Belt drive systems

The simplest arrangement of a belt drive consists of two pulleys connected by the belt as shown in Figure 5.6, where the belt is placed around both pulleys with an initial tension S_0. When a torque T_Q is applied to the input drive shaft in order to drive some load torque T_L at the output drive shaft, it tends to increase the tension on the 'tight' side to S_2 and reduce the tension on the 'slack' side to S_1.

Arrangements such as that shown in Figure 5.6, where the slack side is on top, tend to increase the angle of contact between the belt and pulleys due to gravitational 'sag'. As we will see at a later stage, this is advantageous. Now, if the belt is assumed

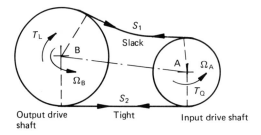

Figure 5.6

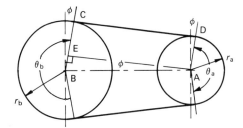

Figure 5.7

to obey Hooke's Law and its total length remains constant, the increase in tension on the tight side $(S_2 - S_0)$ must be equal to the decrease in tension $(S_0 - S_1)$ on the slack side, so

$$S_1 + S_2 = 2S_0 \qquad (5.3)$$

To calculate the nominal length of belt required, we have from Figure 5.7

$$\sin \phi = \frac{BE}{AB} = \frac{r_b - r_a}{AB} \qquad (5.4a)$$

Hence

$$\theta_b = \pi + 2\phi \qquad \theta_a = \pi - 2\phi \qquad (5.4b)$$

thus

$$\text{Belt length} = 2AB \cos \phi + r_b\theta_b + r_a\theta_a \qquad (5.4c)$$

where the angles of 'lap', θ_a and θ_b, are expressed in radians.

There are normally two forms of belt profiles, namely the flat belt (of rectangular cross-section) and the grooved (or V) belt.

5.2.1 Flat-form belts

Consider the free body diagram of the small element of a flat-form belt travelling around, and in contact with, a plain flat pulley of radius r as shown in Figure 5.8.

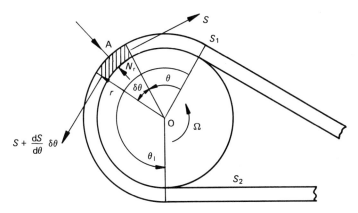

Figure 5.8

For the case where the rotational speed of the pulley is constant at Ω rad/s, then from Section 3.1 the acceleration of the element will have a magnitude $\Omega^2 r$, and act towards the centre of rotation O as shown (centripetal). Therefore, resolving forces parallel to OA and applying Newton's Second Law of Motion, we have

$$S\frac{\delta\theta}{2}+\left(S+\frac{dS}{d\theta}\delta\theta\right)\frac{\delta\theta}{2}-N_r=mr\,\delta\theta\,\Omega^2 r \qquad (5.5)$$

where N_r is the normal force between the pulley and the belt, and m is the mass per length of the belt. It follows therefore since $\delta\theta\,\delta\theta\to 0$ and $\Omega r=V=$ the peripheral speed of the belt, then equation 5.5 reduces to

$$N_r=(S-mV^2)\,\delta\theta \qquad (5.6)$$

Now resolving forces perpendicular to the line OA (neglecting gravitational force), we have, from Section 5.1.1 pertaining to non-slip conditions,

$$\frac{dS}{d\theta}\delta\theta\leqslant\mu_s N_r \qquad (5.7)$$

where μ_s is the coefficient of static friction. Hence combining equations 5.6 and 5.7 we have, for non-slip conditions,

$$\frac{dS}{d\theta}\leqslant\mu_s(S-mV^2)$$

or

$$\frac{dS}{(S-mV^2)}\leqslant\mu_s\,d\theta \qquad (5.8)$$

and integrating over the limits $0<\theta<\theta_1$ and $S_1<S<S_2$, equation 5.8 reduces to

$$\frac{S_2-mV^2}{S_1-mV^2}\leqslant\exp(\mu_s\theta_1) \qquad (5.9)$$

Substituting for S_1 from equation 5.3 we have

$$\frac{S_2 - mV^2}{2S_0 - S_2 - mV^2} \leqslant \exp(\mu_s \theta_1) \tag{5.10}$$

Inspecting equations 5.9 and 5.10 we can see that if μ_s is the same value for both the 'driven' and the 'driver' pulley, slip (if it occurs) will occur at the smaller of the two pulleys, because the angle of lap is less. Additionally we can see that the left-hand side of equation 5.10 is maximum when $V = 0$, i.e. at start-up, whence S_2 is at a maximum; which at the threshold of slip is given by

$$(S_2)_{max} = \frac{2S_0 \exp(\mu_s \theta_1)}{[1 + \exp(\mu_s \theta_1)]} \tag{5.11}$$

The tensile stress due to the above value of tension, plus that due to the enforced curvature of the belt, Ey/r, where E and y are the Young's modulus and half the belt thickness respectively, must not of course be allowed to exceed the maximum permissible tensile stress of the belt as specified by the manufacturer.

Let us now investigate the power that can be transmitted when the belt is on the *threshold of slipping*, i.e.

$$\text{Power} = P = (S_2 - S_1)V = 2(S_2 - S_0)V \tag{5.12}$$

Rearranging equation 5.10 and replacing $\leqslant$ by $=$ gives

$$S_2[1 + \exp(\mu_s \theta_1)] = 2S_0 \exp(\mu_s \theta_1) + mV^2[1 - \exp(\mu_s \theta_1)] \tag{5.13}$$

and combining equations 5.12 and 5.13 gives

$$P = \left(\frac{\exp(\mu_s \theta_1) - 1}{\exp(\mu_s \theta_1) + 1}\right)(2S_0 V - 2mV^3) \tag{5.14}$$

Equation 5.14 describes the power that may just be transmitted at a given peripheral belt speed, V, and any attempts to increase the power would result in slip occurring. Furthermore, for threshold of slip conditions, the speed at which maximum power will be transmitted can be found from

$$\frac{dP}{dV} = 0 = 2S_0 - 6mV'^2$$

i.e.

$V' = $ peripheral speed for maximum power
$$= \sqrt{(S_0/3m)} \tag{5.15}$$

whence

$$P_{max} = 4mV'^3 \frac{\exp(\mu_s \theta_1) - 1}{\exp(\mu_s \theta_1) + 1} \tag{5.16}$$

If a jockey pulley is used on the slack side and is so arranged that S_1 is constant (see Figure 5.9), then $S_0 = S_1$. However, it must be noted that equation 5.3 does not apply in this case.

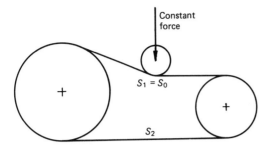

Constant
force

$S_1 = S_0$

S_2

Figure 5.9

Example 5.1

A flat belt 150 mm wide and 8 mm thick passes over a pulley 380 mm in diameter. The angle of lap is 155°. The belt has a mass density of 970 kg/m³. If the maximum permitted belt tensile stress (at start-up) is 2 N/mm², find the belt tensions and the power that could be transmitted at 1200 rev/min with the belt about to slip; $\mu_s = 0.3$ and Young's modulus, $E = 18.2$ N/mm² for the belt material.

For this belt, find the speed for maximum power, the maximum power and the corresponding belt tensions.

SOLUTION

Belt mass per length $= 150 \times 8 \times 970 \times 10^{-6} = 1.164$ kg/m

$\theta_1 = 155° = 2.705$ rad $\Rightarrow$ exp$(\mu_s\theta_1) = 2.2513$

Maximum value of stress due to $S_2 = 2 - (4 \times 18.2/194)$

$$= 1.625 \text{ N/mm}^2$$

therefore

Maximum value of $S_2 = 8 \times 150 \times 1.625$

$$= 1950 \text{ N}$$

(a) *At start-up* $(V = 0)$ and on the threshold of slip we have from equation 5.11

$$1950 = \frac{2S \times 2.2513}{1 + 2.2513} \quad \Rightarrow \quad S_0 = 1408 \text{ N}$$

(b) *At 1200 rev/min* (125.66 rad/s) the peripheral speed of the belt, V, is

$$V = 125.66 \times 0.194 = 24.38 \text{ m/s}$$

hence

$$mV^2 = 1.164 \times (24.38)^2 = 691.8 \text{ N}$$

From equation 5.10, with $=$ replacing $\leqslant$, we have at the threshold of slip $(S_0 = 1408 \text{ N})$:

$$\frac{S_2 - 691.8}{2124.2 - S_2} = 2.2513 \quad \Rightarrow \quad S_2 = 1683.7 \text{ N}$$

and from equation 5.3

$$S_1 = 2 \times 1408 - 1683.7 = 1132.3 \, \text{N}$$

hence

Power transmitted $= (S_2 - S_1)V = 13.44 \, \text{kW}$

(c) *Maximum power condition.* From equation 5.15,

$V' = $ peripheral speed for maximum power

$$= \sqrt{[1408/(3 \times 1.164)]} = 20 \, \text{m/s}$$

and from equation 5.16

$$P_{\max} = 4 \times 1.164 \times 8000 \left(\frac{2.2513 - 1}{2.2513 + 1} \right)$$

$$= 14.34 \, \text{kW at } 984.4 \, \text{rev/min} \, (\equiv 20 \, \text{m/s})$$

$$= 2V'(S_2 - S_0) \Rightarrow S_2 = 1766.5 \, \text{N}$$

and

$$S_1 = 2S_0 - S_2 = 1049.5 \, \text{N}$$

5.2.2 V-form belts

For this type of drive belt the cross-section of the belt is of a form similar to that shown in Figure 5.10a where N'_r is the force normal to the sides of the belt. In this case, therefore, the maximum tangential force which could prevail without slip occurring is $2\mu_s N'_r$, as compared to $\mu_s N_r$ in the case of the flat belt. From Figure 5.10b we deduce that N_r, the normal force parallel to the line OA (as in Figure 5.8), is

$$N_r = 2N'_r \sin \alpha \quad \Rightarrow \quad 2\mu_s N'_r = N_r \mu_s / \sin \alpha \tag{5.17}$$

It follows, therefore, that for the V-form belt the procedures demonstrated and expressions developed in Section 5.2.1 pertaining to flat-form belts are also applicable in this case, except that μ_s is replaced by $\mu_s/\sin \alpha$.

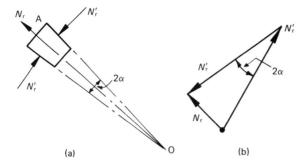

Figure 5.10

5.3   Clutch drive mechanism

A clutch drive mechanism is basically a device for transmitting power between two concentric shafts by means of dry friction. Such mechanisms are extensively used in a wide range of applications—perhaps the most common being the manual gear-operated automobile, where it is normally located between the engine supplying the power and the gearbox as shown schematically in Figure 5.11.

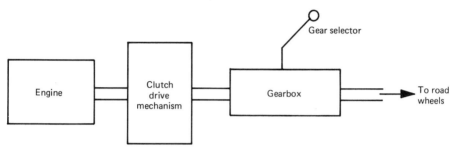

Figure 5.11

The main advantage of such a mechanism is that, by virtue of its design, the power available from the engine crankshaft can be temporarily disconnected from the gearbox input shaft in order to facilitate smooth manual gear changing via the gear selector whilst the engine is still operational and the vehicle is in motion.

Figure 5.12a outlines the basic form of a single-plate clutch drive mechanism as used in manually gear-operated automobiles.

Referring to Figure 5.12a, the smooth-faced flywheel is bolted to a flange fixed on the engine crankshaft and the 'friction plate' is fixed to a boss which is free to slide axially along the gearbox input shaft, but, by means of splines is compelled to rotate with this shaft. Figure 5.12b shows, in more detail, the friction plate which has riveted on each face special friction material annuli. The 'pressure plate', on the other hand, can revolve freely on the gearbox input shaft and is integral with the withdrawal sleeve. A number of spring-loaded fixtures hold the pressure plate to the flywheel and also apply a pressure which keeps the friction plate forcefully sandwiched between the faces of the flywheel and the pressure plate; thus if no slipping occurs total power is transmitted from the crankshaft, through the friction plate, to the gearbox input shaft. If now an axial force is applied to the withdrawal sleeve, having sufficient magnitude to overcome the spring forces at the pressure plate/flywheel connections, the axial pressure on the friction plate will be released, effectively disconnecting the drive between the crackshaft and the gearbox input shaft. The force at the withdrawal sleeve necessary to disengage the friction plate is normally applied by a hydraulic mechanism, activated by depression of the clutch pedal.

Having now described the operation of such a mechanism, let us now proceed to examine the torque and power which can be transmitted by such a mechanism.

5.3.1   Annular contact clutch

This case is identical to that described in Figure 5.12a, where the friction plate is similar to that shown in Figure 5.12b. Consider the case where the mechanism is

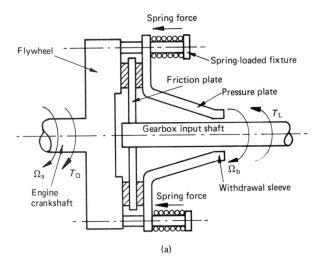

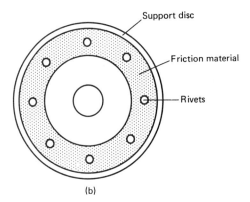

Figure 5.12

engaged such that the double-sided friction plate is subject to an axial pressure, $p(r)$, resulting from the total axial spring force, $F = nf$, where n is the number of springs and f is the force per spring as shown in Figure 5.12a.

Now if slip does not occur, then $\Omega_a = \Omega_b$ and, neglecting losses, we have $T_Q = T_L =$ torque transmitted by the annular friction pads riveted to the friction plate. Consider an annular element of *one* of the friction pads as shown in Figure 5.13 where $p(r)$ is the pressure acting on the element at radius r. The force, δf, normal to this element will therefore be

$$\delta f = 2\pi r \, p(r) \, \delta r \tag{5.18}$$

and the total axial force, F, acting on the complete pad will be given by

$$F = 2\pi \int_{r_1}^{r_2} r \, p(r) \, dr = nf \tag{5.19}$$

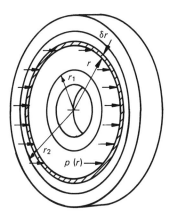

Figure 5.13

Now, if slip is to be avoided, the elemental torque, δT, which can be transmitted by the element under consideration is

$$\delta T \leqslant \mu_s \delta f\, r = 2\pi\mu_s r^2\, p(r)\, \delta r$$

and since there are *two* contacting faces on the friction plate the torque that can be transmitted by the friction plate, T_Q, is

$$T_Q \leqslant 2 \times \left[2\pi\mu_s \int_{r_1}^{r_2} r^2\, p(r)\, dr \right] \tag{5.20}$$

From equations 5.19 and 5.20 we see that the form of $p(r)$ is important in design calculations. Basically there are two criteria, namely:

(1) $p(r) = p = \text{constant}$. This implies that the pressure is constant; in practice this is normally the case for a fairly new friction plate. Hence from equation 5.19 we obtain

$$F = \pi(r_2^2 - r_1^2)p = nf \tag{5.21}$$

and, from equation 5.20,

$$T_Q \leqslant \frac{4\pi\mu_s p(r_2^3 - r_1^3)}{3} = \frac{4\mu_s F[(r_2^3 - r_1^3)/(r_2^2 - r_1^2)]}{3} \tag{5.22}$$

(2) $p(r) = c/r$ where c is a constant. This implies that the pressure decreases linearly from r_1 to r_2 and is termed the *fully worn in* condition. Due to regular engagement and disengagement of the friction plate, it is often found that the wear on the annular friction pads is not uniform but is, instead, approximately proportional to radius, with the result that the pressure applied at some radius (from the total axial thrust, nf) is inversely proportional to the radius. In this case, therefore, substituting in equations 5.19 and 5.20 gives

$$F = nf = 2\pi c(r_2 - r_1) \tag{5.23}$$

and

$$T_Q \leqslant 2\mu_s \pi c(r_2^2 - r_1^2) = \mu_s F(r_2 + r_1) \tag{5.24}$$

Since, for the same value of F, the fully worn in condition renders a lower T_Q than the uniform pressure condition, the pressure distribution associated with the former condition should always be used in design calculations.

5.3.2 Conical contact clutch

Figure 5.14 illustrates the general layout of such a clutch in which the frictional contact area between the flywheel and the friction plate represents the frustum of a cone.

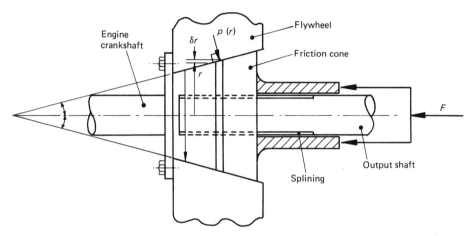

Figure 5.14

The elemental force, δf, acting on the cone in a direction normal to the surface as shown is

$$\delta f = p(r)\, 2\pi r\, \delta r\, \mathrm{cosec}\,\alpha$$

and the axial component of this force, δN, is

$$\delta N = p(r)\, 2\pi r\, \delta r\, \mathrm{cosec}\,\alpha\, \sin\alpha = p(r)\, 2\pi r\, \delta r$$

The total axial force, F, therefore becomes

$$F = 2\pi \int_{r_1}^{r_2} p(r)\, r\, \mathrm{d}r \tag{5.25}$$

Consequently the torque, T_Q, which the cone can transmit without slip occurring is

$$T_Q \leqslant 2\pi\mu_s\, \mathrm{cosec}\,\alpha \int_{r_1}^{r_2} p(r)\, r^2\, \mathrm{d}r \tag{5.26}$$

As in the case of the annular contact clutch, there are two forms of pressure distribution, $p(r)$, to be considered:

(1) $p(r) = p = $ constant. In this case equations 5.25 and 5.26 give

$$F = \pi p(r_2^2 - r_1^2) \tag{5.27a}$$

$$T_Q \leqslant \tfrac{2}{3}\mu_s \pi p(r_2^3 - r_1^3)\operatorname{cosec}\alpha = \tfrac{2}{3}\mu_s F \frac{(r_2^3 - r_1^3)}{(r_2^2 - r_1^2)} \quad \operatorname{cosec}\alpha \tag{5.27b}$$

(2) $p(r) = c/r$, where c is a constant (fully worn in). For this pressure distribution, the corresponding equations for axial force and transmitted torque (under no slip conditions) are

$$F = 2\pi c(r_2 - r_1) \tag{5.28a}$$

$$T_Q \leqslant \pi\mu_s c(r_2^2 - r_1^2)\operatorname{cosec}\alpha = \mu_s \frac{F}{2}(r_2 + r_1)\operatorname{cosec}\alpha \tag{5.28b}$$

Example 5.2

A rear-wheel car is required to be capable of accelerating from 4 to 13 m/s in 3 s while ascending a gradient of 14° in third gear. Estimate a value of constant average torque which the engine must be capable of delivering over the above speed range and a minimum value of clutch axial force in order to ensure that no clutch slip will occur. The vehicle specifications are as follows:

- Total mass of car = 1100 kg.
- Gearbox reduction ratio = 1.36:1.
- Propeller shaft/rear axle reduction ratio = 4.1:1.
- Diameter of road wheels = 0.6 m.
- Clutch type = single plate, double sided, annular.
- Boundary radii of annular friction pads = 70 mm and 108 mm.
- Clutch coefficient of static friction = 0.45.

Neglect the moment of inertia of all rotating parts and assume that road wheel slip does not occur. Gravitational acceleration, $g = 9.81$ m/s.

SOLUTION

Figure 5.15a shows a free body diagram for the car ascending the gradient.

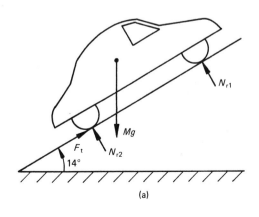

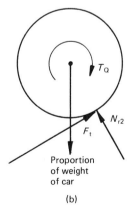

(a)   (b)

Figure 5.15

For the rectilinear motion up the gradient, we have

$$F_t - Mg \sin 14° = M \frac{dv}{dt}$$

i.e.

$$\int_0^3 F_t - 2610.6 \, dt = 1100 \int_4^{13} dv \quad \Rightarrow \quad F_t = 5910.6 \, \text{N}$$

From the free body diagram of the rear wheel, as shown in Figure 5.15b, we have, assuming the moment of inertia to be negligible,

$$T_Q = \text{torque at rear axle} = F_t \times r = 5910.6 \times 0.3 = 1773.2 \, \text{N m}$$

therefore

$$\text{Engine torque} = \frac{T_Q}{1.36 \times 4.1} = 318 \, \text{N m}$$

$$= \text{torque transmitted through clutch}$$

For the clutch mechanism, from equation 5.24, assuming fully worn in conditions,

$$318 \leqslant 0.45 \times F \times 0.178 \quad \Rightarrow \quad F \geqslant 3.97 \, \text{kN}$$

5.4 Friction brake mechanisms

The two most common forms of friction brake mechanism are the shoe brake and disc brake mechanisms.

Figures 5.16a and 5.16b illustrate the basic working principles of an external and internal brake shoe mechanism respectively. In both cases the brake drum is retarded by means of the pressure applied by the stationary (friction lined) brake shoe, initiated

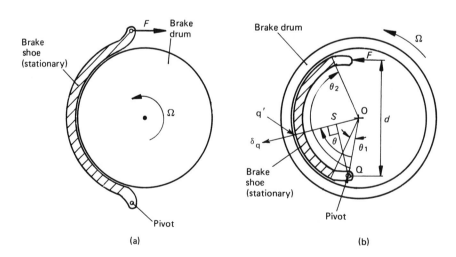

(a)   (b)

Figure 5.16

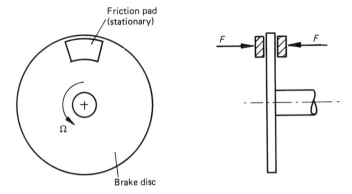

Figure 5.17

by the force F. For both cases shown, any point on the surface of the brake drum will first come into contact with the brake shoe at the point of the shoe nearest to the point of application of the braking force, F. In this situation the brake shoe is termed a 'leading shoe'. Conversely, if the first point of contact is at the point of the shoe nearest to the pivot, i.e. direction of rotation reversed, the brake shoe in this situation would be termed a 'trailing shoe'. As we will show at a later stage, for a given braking force F, a leading shoe system gives more effective braking than a trailing shoe system.

Figure 5.17 outlines the basis of the common disc brake mechanism which comprises a rotating disc and a pair of fixed friction pads. In this case retardation of the brake disc is simply effected by application of a braking force of magnitude F on the friction pads as shown.

For each of these two mechanisms, let us now investigate the braking characteristics.

5.4.1 Shoe brakes

For the sake of convenience we shall analyse the case of the internal shoe mechanism whereupon the reader should verify that the analysis relating to the external shoe mechanism is identical. Furthermore let us consider the leading shoe shown in Figure 5.16b.

We start the analysis by considering the small radial component of the displacement of a point, namely δq, produced by the pivoting of the shoe about the fixed pivot Q. By way of a geometric analysis of the system, it can be shown that δq is proportional to the perpendicular distance from the pivot Q to the radial line Oq', i.e. QS such that

$$\overline{QS} = \overline{OQ} \sin \theta$$

Furthermore, assuming that the friction material attached to the shoe obeys Hooke's Law, the radial pressure, δp, at the point q', acting on the drum, will be proportional to the length QS, i.e.

$$\delta p = K \times \overline{OQ} \times \sin \theta$$

where K is some constant.

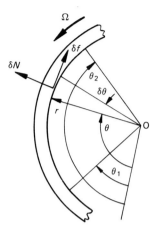

Figure 5.18

Consider now a small element of the brake drum subtended by an angle $\delta\theta$ within the arc of contact bounded by $\theta_1 \leqslant \theta \leqslant \theta_2$ as shown in Figure 5.18.

The force δN acting on the drum in a direction normal to the drum periphery will be given by

$$\delta N = \delta p \times rb\,\delta\theta = Krb\,\overline{OQ}\sin\theta\,\delta\theta$$

where b is the axial width of the brake shoe friction material. Consequently, the tangential friction force, δf, acting on the drum will be

$$\delta f = \mu_k\,\delta N = \mu_k Krb\,\overline{OQ}\sin\theta\,\delta\theta$$

acting in the direction shown, i.e. *opposite* to the direction of the tangential velocity of the drum at that point (see Section 5.1.2). Therefore the total braking torque, T_L (about O), acting on the drum will be

$$T_L = \mu_k Kb\,\overline{OQ}\,r^2 \int_{\theta_1}^{\theta_2} \sin\theta\,d\theta \tag{5.29}$$

Notice that in this case we use μ_k, the coefficient of kinetic friction, since during braking there is relative motion between the drum and the shoe, i.e. slip is occurring. If, however, the drum was stationary, we would substitute μ_s for μ_k and equation 5.29 would then represent the minimum torque necessary to be applied to the drum to overcome static friction, i.e. to initiate movement of the drum past the stationary shoe.

Assuming for the moment that a value of K is known, the braking torque T_L acting on the drum represents a resisting torque on the drum such that, in the absence of any other applied torques, Newton's Second Law of Motion as applied to the drum is

$$- T_L = I_e(d\Omega/dt)$$

where I_e is the dynamic equivalent moment of inertia of the complete system referred to the drum and Ω is the instantaneous rotational speed of the drum.

Consider now the forces acting on the brake shoes, which are shown in Figure 5.19,

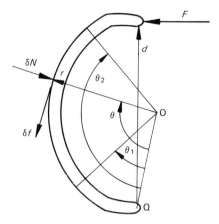

Figure 5.19

where d is the perpendicular distance between Q and the line of action of the braking force, F. Note that, in accordance with Newton's Third Law, the directions of the forces δf and δN are opposite to their direction acting on the drum (see Figure 5.18). Now, since the shoe remains stationary, the total moment of forces about Q must be zero, i.e.

$$Fd = \int_{\theta_1}^{\theta_2} \overline{OQ} \sin \theta \, \delta N - \int_{\theta_1}^{\theta_2} (r - \overline{OQ} \cos \theta) \delta F$$

$$= Krb \, \overline{OQ}^2 \int_{\theta_1}^{\theta_2} \sin^2 \theta \, d\theta - K\mu_k \, \overline{OQ} \, rb \int (r - \overline{OQ} \cos \theta) \sin \theta \, d\theta \qquad (5.30)$$

Equations 5.29 and 5.30 may be utilized in three ways, namely:

(1) For a given system and braking force, F, equation 5.30 can be solved for K which can be substituted into equation 5.29 to find the resulting braking torque, T_L.

(2) For a given system and required braking torque, T_L, equation 5.29 can be solved for K which when substituted in equation 5.30 will render the necessary braking force, F.

(3) For a required braking torque to be produced by a specified braking force, equations 5.29 and 5.30 can be used in conjunction with other design specifications to compute the necessary detail for the design of the system.

Finally, if the direction of rotation of the drum was reversed from that shown in Figure 5.18, i.e. a trailing shoe situation, the reader may wish to verify that equation 5.28 remains unaltered except that the minus sign contained in equation 5.30 becomes a plus sign. The result of this is that for a given braking force, F, equation 5.30 will render a larger value of K for the leading shoe than for the trailing shoe and hence (from equation 5.29) a larger value of braking torque. Figure 5.20 shows the form of the shoe brake mechanism commonly used in automobiles, comprising two identical shoes which are activated by the application of the equal and opposite forces, F, as shown.

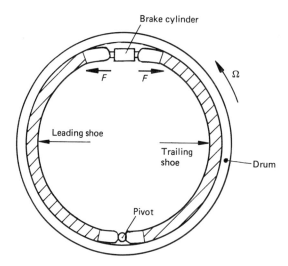

Figure 5.20

Referring to Figure 5.20, it should be noted that the left-hand shoe is a leading shoe whilst the right-hand shoe is a trailing shoe. Therefore denoting K_1 and K_2 as the K values for the left-hand and right-hand shoes respectively, application of equation 5.30 to each shoe gives

$$Fd = K_1 \left\{ rb \, \overline{OQ}^2 \int_{\theta_1}^{\theta_2} \sin^2 \theta \, d\theta - \mu_k \overline{OQ} \, rb \int_{\theta_1}^{\theta_2} (r - \overline{OQ} \cos \theta) \sin \theta \, d\theta \right. \tag{5.31}$$

$$Fd = K_2 \left\{ rb \, \overline{OQ}^2 \int_{\theta_1}^{\theta_2} \sin^2 \theta \, d\theta + \mu_k \overline{OQ} \, rb \int_{\theta_1}^{\theta_2} (r - \overline{OQ} \cos \theta) \sin \theta \, d\theta \right.$$

noting that in the second equation (for the trailing shoe) the minus sign is replaced by a plus as previously explained. Hence, for a given braking force, F, K_1 and K_2 can be calculated from the foregoing equations. Therefore the total braking torque acting on the drum, T_L, becomes (see equation 5.29)

$$T_L = (K_1 + K_2) \mu_k b \, \overline{OQ} \, r^2 \int_{\theta_1}^{\theta_2} \sin \theta \, d\theta \tag{5.32}$$

5.4.2 Disc brakes

Referring to Figure 5.17, consider a sectorial element, of radial width δr, of the disc between the two friction pads, each of which are subjected to the braking force of magnitude F, as shown in Figure 5.21. If the pressure distribution over this element is $p(r)$, the corresponding elemental force, δN, normal to the radial plane of the disc, exerted by this element, is

$$\delta N = p(r) r\theta \, \delta r \tag{5.33}$$

and the frictional torque exerted on the rotating disc, δT_L, is (remembering that there are two pads)

$$\delta T_L = 2\mu_k r \, \delta N = 2\mu_k \theta \, p(r) r^2 \, \delta r \tag{5.34}$$

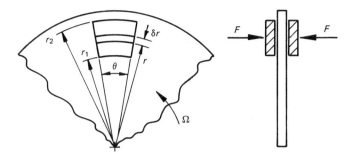

Figure 5.21

Consequently, the total braking torque, T_L, acting on the disc is

$$T_L = 2\mu_k \theta \int_{r_1}^{r_2} p(r) r^2 \, dr \tag{5.35}$$

and by integrating equation 5.33 the braking force, F, is

$$F = \theta \int_{r_1}^{r_2} p(r) r \, dr \tag{5.36}$$

For constant pressure conditions, $p(r) = p$, equations 5.35 and 5.36 become

$$F = \frac{\theta}{2} p(r_2^2 - r_1^2) \tag{5.37a}$$

$$T_L = \tfrac{2}{3} \mu_k \theta p(r_2^3 - r_1^3) = \tfrac{4}{3} F \mu_k \frac{r_2^3 - r_1^3}{r_2^2 - r_1^2} \tag{5.37b}$$

For fully worn in, $p(r) = c/r$,

$$F = \theta c(r_2 - r_1) \tag{5.38a}$$

$$T_L = \mu_k \theta c(r_2^2 - r_1^2) = \mu_k F(r_2 + r_1) \tag{5.38b}$$

For the purpose of design, it is always safest to assume uniform wear conditions.

Problems

5.1 A multiple V belt drive is required to transmit 261 kW at 195 rev/min. The pulley is 1.525 m in diameter and the belt grooves are 40°. The belts have a cross-sectional area of 645 mm², weigh 10.5 N/m and will run with an angle of lap of 160°. The coefficient of static friction may be taken as 0.3.

   If the tensile strength of the belting is 62 N/mm² and a factor of safety of 20 is to be allowed, how many belts will be required? What is the maximum power which this drive could transmit and at what speed would this condition occur?

   Answer: 11 belts; 334.5 kW at 312.6 rev/min.

5.2 A shaft F is connected to a coaxial shaft G by a single plate clutch, with two pairs of friction surfaces whose outer and inner diameters are 120 mm and 70 mm respectively. The total axial load on the clutch is 450 N and the coefficient of static friction is 0.35. Shaft G carries a pinion P gearing with a spur wheel S on a parallel shaft H. The masses and radii of gyration of the three shafts, F, G and H—with attached masses—are 12.5 kg, 80 mm; 20 kg, 70 mm; and 37.5 kg, 120 mm respectively.

Determine the minimum time in which the speed of H can be raised from 500 to 1500 rev/min by a torque applied to shaft F, and the required gear ratio between shafts G and H. Assume fully worn in conditions on the clutch surfaces.

Answer: 3.22 s; 0.426.

5.3 A motor drives a machine through a friction clutch which slips when the torque on it reaches 40 N m. The moment of inertia of the motor armature is 1.6 kg m² and the rotating parts of the machine is 3 kg m². The constant torque developed by the motor is 27 N m and when the clutch is engaged the steady-state speed of the motor and machine is 500 rev/min. At a given instant the clutch is disengaged and remains so for 4 s and then it is re-engaged. Find the time of slipping after re-engagement and determine how much energy is lost during slipping.

Answer: 8.3 s; 17.18 kJ.

5.4 Figure 5.22 shows a friction brake in which the curved lever BC is pivoted at the fixed point C and carries a friction lining which presses on the rotating

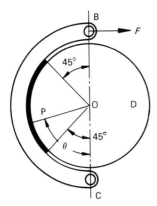

Figure 5.22

drum, D, over an arc of 90°. The diameter of the drum is 250 mm. The distances OB and OC are each 150 mm. The pressure exerted by the friction lining on the drum at any point P is 350 sin θ kN/m². The lining is 50 mm wide and the coefficient of static friction is 0.3. Calculate the braking torque exerted on the drum and the force F to apply the brake. The drum rotates in a clockwise direction.

Answer: 116 N m; 1.02 kN.

6
Out-of-balance and balancing of rotating mass systems

6.1 Introduction

One of the most common forms of machinery in present day use is rotodynamic machinery, usually characterized by a rotating shaft carrying rotors, e.g. turbine or compressor discs. Due to inevitable inaccuracies incurred in the course of manufacture and assembly of such machines, some degree of rotary out-of-balance will always be present, resulting from the centres of mass of the shaft and rotors not lying exactly along the line about which the system is rotating. Although the physical significance of out-of-balance rotating mass systems can be readily appreciated, quantifying the amount by which such a system will be out of balance cannot, in general, be calculated beforehand. However, as we shall see in Section 6.4, simple tests carried out on a rotodynamic system can readily quantify and describe the amount by which the system is out of balance.

6.2 Out-of-balance forces and moments

Consider a single rotor mounted on a rotating shaft as shown in Figure 6.1a. If the rotational speed is maintained constant, the out-of-balance force acting on the shaft can be modelled as the centrifugal force (the inertia force associated with the centripetal acceleration, i.e. a force of magnitude $m\Omega^2 r$ acting in a direction radially away from the centre of rotation) associated with some small mass, m, at some radius r and angularly positioned at some angle β from the rotating reference line. More often, in cases such as this, the centrifugal force vector is termed the frame force vector, **FF**, such that

$$\mathbf{FF} = m\Omega^2 r[-\mathbf{i}\sin(\Omega t + \beta) + \mathbf{j}\cos(\Omega t + \beta)]$$

where **i** and **j** are unit vectors along the fixed reference axes OX and OY respectively, and the term $m\Omega^2 r$ represents the magnitude of the centrifugal force. Now at this stage, in an effort to simplify future analysis, it is beneficial to express this frame force vector in complex variable notation thus:

$$\mathbf{FF} = \Omega^2 \vec{C}(-\mathbf{i}\sin\Omega t + \mathbf{j}\cos\Omega t) \tag{6.1}$$

where $\vec{C}$ is the complex factor of $\mathbf{FF} = mr\underline{/\beta}$ and the symbol $\underline{/\beta}$ denotes the phase angle with reference to the rotating reference line.

If we now place this same rotor at some axial distance z from O, along the axis

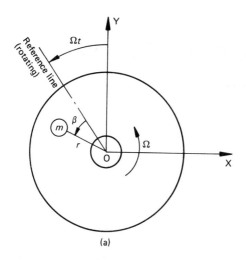

(a)

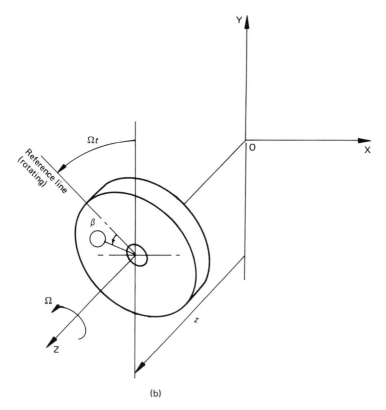

(b)

Figure 6.1

OZ as shown in Figure 6.1b, the frame force moment vector, **MM**, about O is

$$\mathbf{MM} = \mathbf{k}z \wedge \Omega^2 \vec{C}(-\mathbf{i}\sin\Omega t + \mathbf{j}\cos\Omega t)$$

where **k** is unit vector along OZ and the symbol $\wedge$ denotes vector cross multiplication. Therefore

$$\mathbf{MM} = -\Omega^2 z\vec{C}(\mathbf{i}\cos\Omega t + \mathbf{j}\sin\Omega t) \tag{6.2}$$

Likewise if we were to place n rotors on the shaft, each having their own specific values of m, r and β, the frame force and frame force moment vectors become

$$\mathbf{FF}/\Omega^2 = \left(\sum_{j=1}^{n} \vec{C}_j\right)(-\mathbf{i}\sin\Omega t + \mathbf{j}\cos\Omega t) \tag{6.3a}$$

$$\mathbf{MM}/\Omega^2 = -\left(\sum_{j=1}^{n} z_j\vec{C}_j\right)(\mathbf{i}\cos\Omega t + \mathbf{j}\sin\Omega t) \tag{6.3b}$$

where in equation 6.3b care should be exercised in applying the correct sign to each value of z_j, i.e. plus if it is in the same direction as the axis OZ and minus if it is not.

Therefore, assuming that such systems are supported by bearings, the sum of all support bearing reaction force vectors *acting on the shaft*, **RF**, and the frame force vector must be zero, i.e.

$$\mathbf{FF} + \mathbf{RF} = 0 \tag{6.4a}$$

Likewise the sum of all support bearing force moment vectors acting on the shaft, **RM**, and the frame force moment vector must also be zero, i.e.

$$\mathbf{MM} + \mathbf{RM} = 0 \tag{6.4b}$$

Example 6.1

A rotodynamic machine consists of three impellers mounted on a shaft, which is supported by two bearings at A and B as shown in Figure 6.2. From individual tests

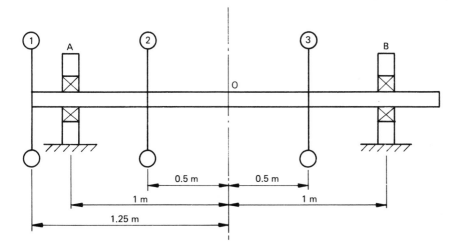

Figure 6.2

carried out on each impeller, the respective values of the product mr and angle β (with respect to a reference line on the shaft) were found to be as follows:

Line 1 $mr/(\text{kg m}) = 2.4 \times 10^{-4}$ $\beta = +30°$

Line 2 $mr/(\text{kg m}) = 7 \times 10^{-4}$ $\beta = -30°$

Line 3 $mr/(\text{kg m}) = 3.5 \times 10^{-4}$ $\beta = -120°$

Describe the form of the bearing reaction forces.

SOLUTION

$$\sum_{j=1}^{3} \vec{C}_j = \sum_{j=1}^{3} (mr)_j \underline{/\beta_j}$$

$$= [2.4\underline{/30°} + 7\underline{/-30°} + 3.5\underline{/-120°}] \times 10^{-4}$$

$$= [2.4(\cos 30° + i\sin 30°) + 7(\cos 30° - i\sin 30°)$$

$$\quad + 3.5(\cos 120° - i\sin 120°)] \times 10^{-4}$$

$$= (6.3907 - i\,5.331) \times 10^{-4}$$

$$= 8.322 \times 10^{-4} \underline{/-39.8°}$$

where i is the complex conjugate $\sqrt{-1}$.

Similarly, with respect to the point O in Figure 6.2,

$$\sum_{j=1}^{3} z_j \vec{C}_j = [(+1.25)2.4\underline{/30°} + (0.5)7\underline{/-30°} + (-0.5)3.5\underline{/-120°}] \times 10^{-4}$$

$$= 6.626 \times 10^{-4} \underline{/11°}$$

Hence from equations 6.3a and 6.3b we have

$$\mathbf{FF}/\Omega^2 = (8.322 \times 10^{-4} \underline{/-39.8°})(-i\sin\Omega t + j\cos\Omega t) \tag{1}$$

and

$$\mathbf{MM}/\Omega^2 = -(6.626 \times 10^{-4} \underline{/11°})(i\cos\Omega t + j\sin\Omega t) \tag{2}$$

Now let us describe the bearing reaction forces on the shaft at A and B as $\mathbf{RF_A}$ and $\mathbf{RF_B}$ respectively, which in general terms may be written as

$$\mathbf{RF_A}/\Omega^2 = \vec{C}_A(-i\sin\Omega t + j\cos\Omega t)$$
$$\mathbf{RF_B}/\Omega^2 = \vec{C}_B(-i\sin\Omega t + j\cos\Omega t)$$

and the total bearing reaction force on the shaft, $\mathbf{RF}$, is

$$\mathbf{RF} = \mathbf{RF_A} + \mathbf{RF_B} \tag{3}$$

Hence from equation 6.4a, 1 and 3 we have

$$\vec{C}_A + \vec{C}_B = -8.322 \times 10^{-4} \underline{/-39.8°} \tag{a}$$

Also the moment about O of the bearing reaction force vectors, $\mathbf{RM}$, is

$$\mathbf{RM}/\Omega^2 = -[(+1)\vec{C}_A + (-1)\vec{C}_B][i\cos\Omega t + j\sin\Omega t] \tag{4}$$

Hence from equations 6.4b, 2 and 3 we have

$$\vec{C}_A - \vec{C}_B = -6.626 \times 10^{-4} \underline{/11°} \tag{b}$$

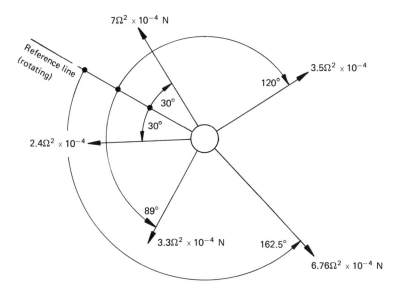

Figure 6.3

and solving for C_A and C_B from (a) and (b) gives

$$\vec{C}_A = 6.76 \times 10^{-4} \underline{/162.5°} \qquad \vec{C}_B = 3.3 \times 10^{-4} \underline{/89°}$$

Therefore, as shown in Figure 6.3, $\mathbf{RF}_A$ has a magnitude of $6.76\Omega^2 \times 10^{-4}\,\mathrm{N}$ and *leads* the rotating reference line by 162.5°, whilst $\mathbf{RF}_B$ has a magnitude of $3.3\Omega^2 \times 10^{-4}\,\mathrm{N}$ and leads the reference line by 89°.

Now since $\mathbf{RF}_A$ and $\mathbf{RF}_B$ are the reaction force vectors *applied to the shaft* by the bearings at the lines A and B respectively in Figure 6.2, then, by Newton's Third Law, the reaction force vectors *applied to the bearings* by the shaft at A and B will be equal in magnitude but opposite in direction to the vectors $\mathbf{RF}_A$ and $\mathbf{RF}_B$. From this statement, therefore, we can deduce that the out-of-balance force vectors at lines 1, 2 and 3 can be represented, and thus replaced, by two force vectors, $-\mathbf{RF}_A$ and $-\mathbf{RF}_B$, at planes A and B respectively. As we shall see at a later stage in this chapter, replacing the total out-of-balance frame force vector by force vectors at any two planes lends itself ideally to the exercise of balancing.

6.3   Balancing of frame forces and frame force moments

Consider once again the single rotor system shown in Figures 6.1a and 6.1b. For such a system, if we know values for β and the product mr, complete balancing could be achieved simply by placing *on* the rotor a balance mass, m_b, at some radius r_b as shown in Figure 6.4a, such that $mr = m_b r_b$. This is an example of *single plane balancing* and is useful in cases where there is only one out-of-balance rotor and it is practically permissible to place the balance mass *on* the rotor. If, however, we were to place the balance mass at the same radius, r_b, and at the same orientation to the rotating reference line, but at some axial distance from the rotor as in Figure 6.4b, then although the frame force vector **FF** is still zero, by virtue of the axial displacement between the rotor and m_b, a frame force moment vector, **MM**, about O, will be present, i.e. complete balancing has not been achieved. In order to achieve complete balancing, if it is not permissible to place masses on the rotor, we would be required to place suitably valued balance masses (of different values from m_b above) at suitable radii on two separate planes as shown in Figure 6.4c.

We now turn our attention to a system containing more than one rotor where each rotor has its own individual out-of-balance force. For such a system, if it is possible to either calculate or experimentally determine the out-of-balance force

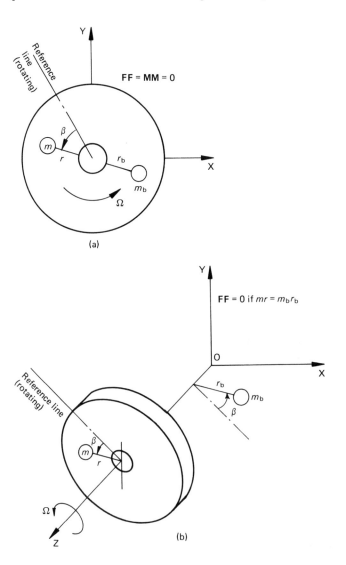

Figure 6.4

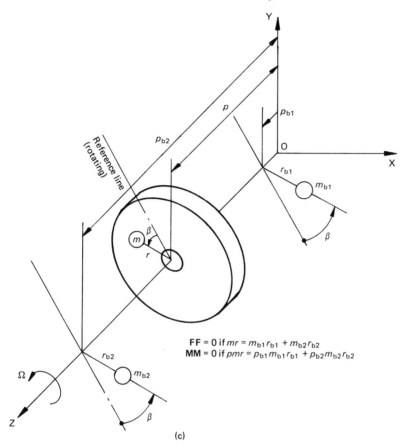

$$FF = 0 \text{ if } mr = m_{b1}r_{b1} + m_{b2}r_{b2}$$
$$MM = 0 \text{ if } pmr = p_{b1}m_{b1}r_{b1} + p_{b2}m_{b2}r_{b2}$$

(c)

Figure 6.4 (*Cont.*)

associated with each of the rotors, *and* it is practically permissible to add balance masses to each rotor, complete balancing of the system can be achieved by applying single plane balancing (as in Figure 6.4a) at each of the rotors. In practice, however, such a means of obtaining complete balance would be impractical. Alternatively, the reader will recall that at the end of Section 6.2 we deduced that the total frame force vector due to any number of out-of-balance rotors may be re-expressed in terms of force vectors acting in any two planes located along the axis of rotation. In this way, if it is possible to determine the form of these force vectors in each of the two planes, single plane balancing can be applied in the same two planes. In an attempt to illustrate this, let us once again consider the system described in Example 6.1 with the requirement that total balance is to be achieved by adding balance masses at planes D and E as shown in Figure 6.5.

From equations 1 and 2 of Example 6.1, we have the frame force and frame force moment vector (about O) due to the rotors as

$$\mathbf{FF}/\Omega^2 = (8.322 \times 10^{-4}\underline{/-39.8°})(-\mathbf{i}\sin\Omega t + \mathbf{j}\cos\Omega t) \tag{1}$$

$$\mathbf{MM}/\Omega^2 = -(6.626 \times 10^{-4}\underline{/11°})(\mathbf{i}\cos\Omega t + \mathbf{j}\sin\Omega t) \tag{2}$$

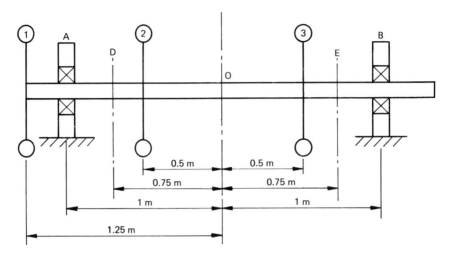

Figure 6.5

and we wish to replace the vector **FF** by two force vectors at planes D and E, i.e. $\mathbf{F_D}$ and $\mathbf{F_E}$ respectively. We describe these forces in general terms thus:

$$F_D/\Omega^2 = \vec{C}_D(-\mathbf{i}\sin\Omega t + \mathbf{j}\cos\Omega t) \tag{3}$$

and

$$F_E/\Omega^2 = \vec{C}_E(-\mathbf{i}\sin\Omega t + \mathbf{j}\cos\Omega t) \tag{4}$$

Hence, equating **FF** to the sum of $\mathbf{F_D}$ and $\mathbf{F_E}$ gives

$$\vec{C}_D + \vec{C}_E = 8.322 \times 10^{-4}\underline{/-39.8^\circ} \tag{a}$$

Similarly the moments of $\mathbf{F_D}$ and $\mathbf{F_E}$ about O must equal **MM**, giving

$$0.75\vec{C}_D - 0.75\vec{C}_E = 6.626 \times 10^{-4}\underline{/11^\circ} \tag{b}$$

Solving equations (a) and (b) above gives

$$\vec{C}_D = 7.75 \times 10^{-4}\underline{/-13.6^\circ} \qquad \vec{C}_E = 3.687 \times 10^{-4}\underline{/-108^\circ}$$

Figures 6.6a and 6.6b illustrate the forces $\mathbf{F_D}$ and $\mathbf{F_E}$ respectively relative to the common rotating reference line. Hence, $\mathbf{F_D}$ and $\mathbf{F_E}$ can be balanced by introducing at planes D and E, respectively, the balance masses m_{b1} and m_{b2} at radii r_1 and r_2 as shown. Thus if r_1 and r_2 are specified (as they normally are) the values of m_{b1} and m_{b2} can be calculated from the expressions

$$m_{b1} = |\vec{C}_D|/r_1 \qquad m_{b2} = |\vec{C}_E|/r_2$$

In the above example, the reader will appreciate that in order to determine the value and position of the balance masses in planes D and E it was necessary for us to know the magnitude and direction of the out-of-balance force vectors at these planes. In practice, these force vectors can really only be determined experimentally. In Section 6.4, therefore, we shall demonstrate how such force vectors may be obtained.

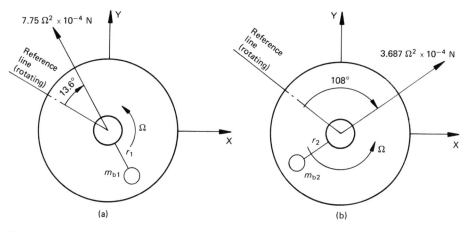

Figure 6.6

6.4   Experimental method for balancing of rotodynamic machinery

In the previous sections of this chapter we considered the theory surrounding the out-of-balance and balancing of rotodynamic machinery. Basically, if out-of-balance forces are left unabated then, as we have seen, they will give rise to harmonic reaction forces at the bearings supporting the machinery. Subsequently these forces will give rise to harmonic vibration at the bearings and the net effect of this is to greatly reduce the working life of the bearings. However, common sense would indicate to us that the level of the vibration at the bearings, irrespective of how we express it, will in some way be a measure of the intensity of the total out-of-balance force acting on the system.

Consider the system shown in Figure 6.7a where a rotor (or series of rotors) is rotating with an angular velocity of Ω rad/s. This will inevitably give rise to some harmonic force vector at one of the support bearings shown.

Now due to this force some harmonic vibratory response will be produced at the bearing, and this can be detected by attaching to the bearing a vibration level indicator. The most common type of indicator is an accelerometer, the output from which will be a harmonic trace at a frequency corresponding to the frequency of rotation of the shaft, and having an amplitude which will in some way be related to the amplitude of the reaction force vector acting on that bearing. Simultaneously let us assume that we can detect some rotating reference line as it passes a fixed point as shown—this normally being achieved by either a photocell or electromagnetic sensor device. If we were now to pass both signals to an analyser, we would expect to see two signals (reference line indicator and vibration indicator) separated by some phase ϕ as shown in Figure 6.7b. This phase is however a combination of the phase between the reference line and the reaction force vector, the inherent phase shifts associated with the instrumentation, and the angular positional phase between the vibration and the reference line indicators. Furthermore, it should be noted that the contribution due to the phase shift associated with the instrumentation is most often not constant, but

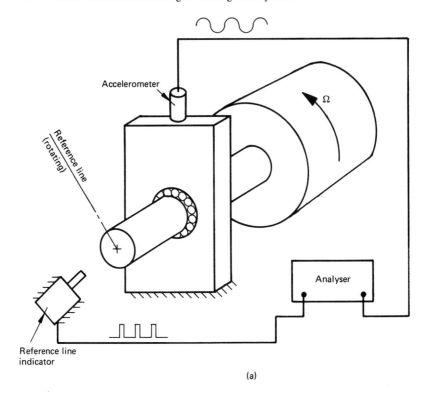

(a)

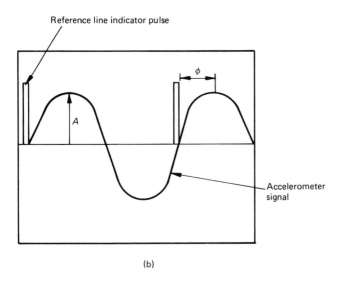

(b)

Figure 6.7

some function of the frequency Ω. Therefore all we can assume for the moment is that, with respect to the complex factor of the vibration trace, $\vec{V}(= A\underline{/\phi})$, its amplitude and phase are in some way related to the amplitude of the force vector acting on the bearing and its phase relative to the reference marker. Now, if we denote the complex factor of the force vector acting on the bearing as $\vec{C}$, we can write

$$\vec{V} = \vec{T}\vec{C} \tag{6.5}$$

where $\vec{T}$ is a transfer operator relating the complex factor of the vibration trace to $\vec{C}$ for *a given speed*. Let us now proceed to demonstrate how this important relationship can be applied to the balancing of rotodynamic machinery.

6.4.1 Single-plane balancing

Consider the system shown in Figure 6.8 where a single rotor driven by a motor is supported by a single bearing and it is assumed that the only out-of-balance force present is that due to the rotor alone.

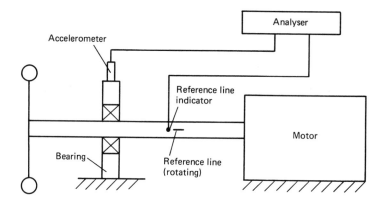

Figure 6.8

Now when the system is running at some speed Ω we can write as in equation 6.5

$$\vec{V} = \vec{T}\vec{C} \tag{6.6}$$

where the complex factor of the vibration trace, $\vec{V}$, will be obtained from the analyser, i.e. $A\underline{/\phi}$, where A is its amplitude and ϕ is its phase relative to the reference indicator pulse. Note, however, that $\vec{T}$ and $\vec{C}$ are unknown at this stage. If we now stop the machine and place on the rotor some arbitrary mass, m, at some arbitrary radius, r, and some arbitrary angle relative to the reference line, β, then, when the machine is run at the same speed as before in the same direction, the vibration trace will change accordingly to $\vec{V}'$, such that

$$\vec{V}' = \vec{T}(\vec{C} + mr\underline{/\beta}) \tag{6.7}$$

Therefore, from equations 6.6 and 6.7 we can solve for $\vec{C}$ and $\vec{T}$, and, since $\vec{C}$ is directly related to the out-of-balance force at the rotor, then having calculated it we can proceed to apply single plane balancing.

Example 6.2

For the system described in Figure 6.8, the amplitude of the vibration trace at the analyser was three units and its phase relative to the reference line indicator pulse was 10°. The machine was stopped and a trial mass of 0.1 kg was placed on the rotor at a radius of 5 cm and at an angle of − 30° from the reference line, i.e. 30° measured in a direction opposite to the direction of rotation. When the machine was re-run at the same speed as before (and in the same direction), the amplitude of the vibration trace was 4.5 units and its phase was 120°. Calculate the necessary value of a balance mass to be attached to the rotor at a radius of 5 cm and specify its position relative to the reference line on the rotor.

SOLUTION

Without the trial mass and from equation 6.6 we have

$$3\underline{/10°} = \vec{T}\,\vec{C} \tag{1}$$

With the trial mass and from equation 6.7 we have

$$4.5\underline{/120°} = \vec{T}(\vec{C} + 0.005\underline{/-30°}) \tag{2}$$

Hence from equations 1 and 2 we have

$$\vec{T} = 1243.57\underline{/117°} \qquad \vec{C} = 0.0024\underline{/-167°}$$

Figure 6.9 shows the rotor complete with reference line, $\vec{C}$, and the required angular position of the balance mass.
Hence $0.0024 = m_b r_b \Rightarrow m_b = 0.0024/0.05 = 0.048\,\text{kg}.$

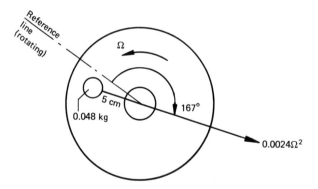

Figure 6.9

6.4.2 Two-plane balancing

In Section 6.3 we considered the means by which a multi-rotor system could be balanced, namely by replacing the combined out-of-balance frame force vector of all the individual rotors by force vectors at two separate planes. Having determined the magnitude and phase of these two forces, we could subsequently apply single plane balancing at each of these planes. Therefore, in practice, if it were possible to establish, experimentally, the magnitude and phase of the forces acting at any two planes—such forces representing the total out-of-balance force due to the rotors—single plane balancing could be duly implemented by the addition of balance masses at the correct radial and angular positions in these planes. Consider the system shown in Figure 6.10 comprising a shaft carrying three rotors supported by the two bearings at planes A and B.

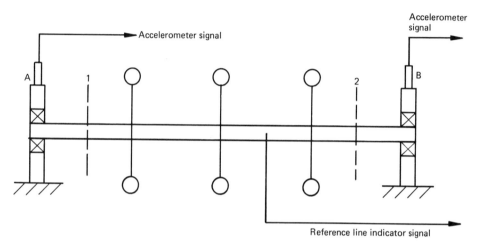

Figure 6.10

As in the case of single plane balancing, we have a reference line indicator and vibration transducers mounted on each of the bearings, with all three signals passed to an analyser as before. Let planes 1 and 2 be the planes at which we require to apply the balancing. Now if we denote $\vec{V}_A$ and $\vec{V}_B$ as the components of the vibration traces for bearings A and B respectively, then for bearing A we can write

$$\vec{V}_A = \vec{T}_{A1}\vec{C}_1 + \vec{T}_{A2}\vec{C}_2 \tag{6.8}$$

where $\vec{C}_1$ and $\vec{C}_2$ are the complex factors of the force vectors acting on planes 1 and 2 respectively, and $\vec{T}_{A1}$ and $\vec{T}_{A2}$ are transfer operators which relate $\vec{V}_A$ to $\vec{C}_1$ and $\vec{C}_2$ respectively. Similarly for bearing B we have

$$\vec{V}_B = \vec{T}_{B1}\vec{C}_1 + \vec{T}_{B2}\vec{C}_2 \tag{6.9}$$

If we now stop the system and add at plane 1 a trial mass m_{t1} at some arbitrary radius r_{t1} and phase relative to the reference line, β_1, and then run the system again at the same speed as before and in the same direction, the vibration traces from A and B will change to $\vec{V}'_A$ and $\vec{V}'_B$, such that

$$\vec{V}'_A = \vec{T}_{A1}(\vec{C}_1 + m_{t1}r_{t1}\underline{/\beta_1}) + \vec{T}_{A2}\vec{C}_2 \tag{6.10}$$

$$\vec{V}'_B = \vec{T}_{B1}(\vec{C}_1 + m_{t1}r_{t1}\underline{/\beta_1}) + \vec{T}_{B2}\vec{C}_2 \tag{6.11}$$

Let us once again stop the system, remove the trial mass from plane 1, and add at plane 2 a trial mass m_{t2} at a radius r_{t2} and angle β_2 to the reference line. Upon running the system once again, at the same speed and in the same direction, the vibration traces from A and B will change to $\vec{V}''_A$ and $\vec{V}''_B$, such that

$$\vec{V}''_A = \vec{T}_{A1}\vec{C}_1 + \vec{T}_{A2}(\vec{C}_2 + m_{t2}r_{t2}/\beta_2) \tag{6.12}$$

$$\vec{V}''_B = \vec{T}_{B1}\vec{C}_1 + \vec{T}_{B2}(\vec{C}_2 + m_{t2}r_{t2}/\beta_2) \tag{6.13}$$

Subtracting equation 6.8 from equation 6.10 and equation 6.9 from equation 6.11 gives

$$\vec{T}_{A1} = \frac{\vec{V}'_A - \vec{V}_A}{m_{t1}r_{t1}/\beta_1} \tag{6.14a}$$

$$\vec{T}_{B1} = \frac{\vec{V}'_B - \vec{V}_B}{m_{t1}r_{t1}/\beta_1} \tag{6.14b}$$

Subtracting equation 6.8 from 6.12 and equation 6.9 from 6.13 gives

$$\vec{T}_{A2} = \frac{\vec{V}''_A - \vec{V}_A}{m_{t2}r_{t2}/\beta_2} \tag{6.15a}$$

$$\vec{T}_{B2} = \frac{\vec{V}''_B - \vec{V}_B}{m_{t2}r_{t2}/\beta_2} \tag{6.15b}$$

Consequently, from equations 6.8 and 6.9 we obtain

$$\vec{C}_1 = \frac{\vec{T}_{B2}\vec{V}_A - \vec{T}_{A2}\vec{V}_B}{\vec{T}_{A1}\vec{T}_{B2} - \vec{T}_{A2}\vec{T}_{B1}} \tag{16.16a}$$

$$\vec{C}_2 = \frac{\vec{T}_{B1}\vec{V}_A - \vec{T}_{A1}\vec{V}_B}{\vec{T}_{B1}\vec{T}_{A2} - \vec{T}_{A1}\vec{T}_{B2}} \tag{16.16b}$$

Having determined $\vec{C}_1$ and $\vec{C}_2$ we now proceed to apply single plane balancing at planes 1 and 2 in the normal manner.

Example 6.3

For the system shown in Figure 6.10 the vibratory amplitudes and phases (relative to the reference line indicator pulse) shown in Table 6.1 were found to exist at the bearings A and B for various stages of a test to determine the out-of-balance forces acting at planes 1 and 2. The trial mass in both cases had a value of 0.1 kg, set at a radius of 5 cm, and at an angle of 30° from the reference line, measured in a direction *opposite* to that of the direction of rotation. Determine the value and position of balance masses required to be placed at planes 1 and 2, at a radius of 5 cm, in order to render the machine in a state of complete balance.

Table 6.1

	Vibratory amplitude (units)		Phase (degrees)	
	A	B	A	B
Machine alone	3	3.5	10	15
Balance mass at 1	4	3	16	30
Balance mass at 2	0.5	1.2	120	110

SOLUTION

$$\vec{V}_A = 3\underline{/10^\circ} \qquad \vec{V}'_A = 4\underline{/16^\circ} \qquad \vec{V}''_A = 0.5\underline{/120^\circ}$$
$$V_B = 3.5\underline{/15^\circ} \qquad V'_B = 3\underline{/30^\circ} \qquad V''_B = 1.2\underline{/110^\circ}$$

From equations 6.14a, 6.14b, 6.15a and 6.15b we have

$$\vec{T}_{A1} = \frac{4\underline{/16^\circ} - 3\underline{/10^\circ}}{0.005\underline{/-30^\circ}} = 212.74\underline{/63.14^\circ}$$

$$\vec{T}_{B1} = \frac{3\underline{/30^\circ} - 3.5\underline{/15^\circ}}{0.005\underline{/-30^\circ}} = 196.53\underline{/-7.2^\circ}$$

$$\vec{T}_{A2} = \frac{0.5\underline{/120^\circ} - 3\underline{/10^\circ}}{0.005\underline{/-30^\circ}} = 641.125\underline{/31.57^\circ}$$

$$\vec{T}_{B2} = \frac{1.2\underline{/110^\circ} - 3.5\underline{/15^\circ}}{0.005\underline{/-30^\circ}} = 759.5\underline{/26.64^\circ}$$

Therefore from equations 16.16a and 16.16b we have

$$\vec{C}_1 = 0.00248\underline{/0^\circ} \qquad \vec{C}_2 = 0.0052\underline{/-14.3^\circ}$$

Therefore

$$m_{b1} = \frac{|\vec{C}_1|}{0.05} = 0.0497 \, \text{kg at } 180^\circ$$

as shown in Figure 6.11a, and

$$m_{b2} = \frac{|\vec{C}_2|}{0.05} = 0.104 \, \text{kg at } 194.3^\circ$$

as shown in Figure 6.11b.

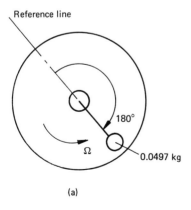

(a)

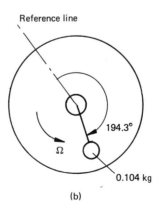

(b)

Figure 6.11

Problems

6.1 A three-bladed propeller has blades A, B and C which weigh 93.8, 94.2 and
94.6 N respectively. The positions of the centroids of the three blades are 1.024,
1.026 and 1.022 m respectively from the propeller centreline. The relative angular
positions of the blades are shown in Figure 6.12.

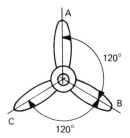

Figure 6.12

Determine the magnitude and direction of the out-of-balance force at 1800
rev/min.

Answer: 2229.44 N at 177.6° anticlockwise from A.

6.2 A shaft is supported in bearings 1.2 m apart and carries three pulleys in planes
A, B and C as shown in Figure 6.13. The pulleys are all out-of-balance, the
amounts (*mr*) being 0.1, 0.15 and 0.15 respectively. However, the shaft is 'statically
balanced' in this condition, i.e. the frame force vector **FF** is zero.

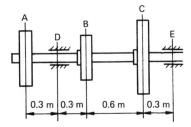

Figure 6.13

Determine the magnitude of the bearing reactions when the shaft rotates at
500 rev/min.

Answer: 282.5 N.

6.3 For the system shown in Figure 6.8, it is required to apply single plane balancing
at the rotor. Under normal operating conditions it is found that the trace of
the vibration at the bearing had an amplitude of 4 units and a phase of 8° with
respect to the reference line indicator pulse. The machine was stopped and a

trial mass of value 0.1 kg was placed on the rotor at a radius of 5 cm and at an angle of 30° to the reference line, measured in a direction *opposite* to that of the normal direction of rotation of the shaft. Subsequently when the machine was run again under the same conditions, the vibratory trace amplitude was found to be 6 units at a phase of 180°.

Calculate the necessary value of a balance mass to be fixed on the rotor at a radius of 5 cm and specify its angular position relative to the reference line marked on the rotor.

Comment on the following points relating to *in situ* balancing:

(a) the consequence if the machine is run at differing speeds with and without the trial mass,

(b) the reference line which activates the reference line indicator is different from the reference line marked on the rotor.

Answer: 0.04 kg at 143° anticlockwise.

7

Out-of-balance and balancing of reciprocating mass systems

7.1 Introduction

Consider the frictionless drive mechanism shown in Figure 7.1 where the crankshaft, O, is rotating at constant angular velocity, Ω rad/s, in the sense indicated.

The displacement vector, $\mathbf{x}$, of the piston, P, from the crankshaft, O, is such that

$$\mathbf{x} = \mathbf{i}(r\cos\theta + l\cos\phi) \tag{7.1}$$

where $\mathbf{i}$ is unit vector along the horizontal. Noting that $l\sin\phi = r\sin\theta$, then

$$\cos\phi = (1 - c^2\sin^2\theta) \qquad \text{where } c = r/l \tag{7.2}$$

Applying the Taylor expansion theorem to equation 7.2 and noting that

$$\sin\theta = [\exp(i\theta) - \exp(-i\theta)]/2 \tag{7.3a}$$

and

$$\cos = [\exp(i\theta) + \exp(-i\theta)]/2 \tag{7.3b}$$

where i is the complex conjugate, $\sqrt{-1}$, then after some algebraic manipulation equations 7.1, 7.2, 7.3a and 7.3b combine to give

$$\mathbf{x} = \mathbf{i}r\left(A_1\cos\Omega t + \frac{A_2}{4}\cos 2\Omega t + \frac{A_4}{16}\cos 4\Omega t + \frac{A_6}{36}\cos 6\Omega t + \text{higher even powers}\right) \tag{7.4}$$

where $\Omega t = \theta$ and

$$\begin{aligned}
A_1 &= 1 \\
A_2 &= c + \tfrac{1}{4}c^3 + \tfrac{15}{128}c^5 + \text{higher odd powers} \\
A_4 &= -\tfrac{1}{4}c^3 - \tfrac{3}{16}c^5 + \text{higher odd powers} \\
A_6 &= \tfrac{9}{128}c^5 + \text{higher odd powers}
\end{aligned} \tag{7.5}$$

Note that, since in practical engine configurations the ratio $c \doteqdot \tfrac{1}{3}$, we can truncate equations 7.4 and 7.5 at all terms containing c^5, or even before. Therefore the acceleration vector, $\mathbf{a}$, of the piston is

$$\mathbf{a} = \mathrm{d}^2\mathbf{x}/\mathrm{d}t^2 = -i\Omega^2 r(A_1\cos\Omega t + A_2\cos 2\Omega t + A_4\cos 4\Omega t + A_6\cos 6\Omega t) \tag{7.6}$$

and the inertia force vector acting on the piston will be $-M\mathbf{a}$, where M is the total reciprocating mass at the piston. This will be discussed at a later stage. Assuming for the present that all other components of the mechanism are massless and neglecting

100

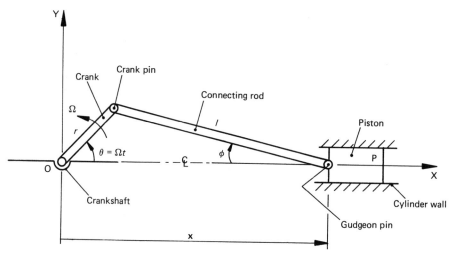

Figure 7.1

the gas pressure force on the piston, the inertia force vector described will also be the force acting on the crankshaft bearings, commonly termed the *frame force vector*, **FF**, which is described more explicitly as

$$\mathbf{FF} = \sum \mathbf{FF}_n \qquad n = 1, 2, 4, 6, \text{etc.} \tag{7.7}$$

where $\mathbf{FF}_n$ is the nth order harmonic frame force vector given by

$$\mathbf{FF}_n = i\Omega^2 C_n \cos n\Omega t \tag{7.8}$$

where

$$C_n = MrA_n \tag{7.9}$$

and expressions for values of A_n are listed in equation 7.5. Unlike the out-of-balance rotating forces described in Chapter 6, the nth order out-of-balance forces in this study act only along a straight line corresponding to the line of action of the reciprocation of the mass, M, i.e. along the centreline indicated in Figure 7.1.

The reader will appreciate, from the examples considered in Chapter 3, that any mass associated with the connecting rod will result in an additional force at the crankshaft bearings. In order to allow for this force, the mass of the connecting rod can be divided (dynamically) into two components, namely one component positioned at the crank pin and another at the gudgeon pin. Considering the mass distribution of a typical connecting rod (Figure 7.2a) this would seem reasonable. Thus the reciprocating mass M in equation 7.9 is the sum of the piston mass and a component of the connecting rod mass, and is termed the **total effective reciprocating mass**. The out-of-balance contribution from the crank combined with the referred mass of the connecting rod is usually cancelled by means of careful design of the volume profile of the crank as can be seen from Figure 7.2b which shows a typical engine crank profile.

We can therefore continue the analysis on the basis that the only out-of-balance forces are those due to the reciprocating nth order frame forces, $\mathbf{FF}_n$, described by equation 7.8.

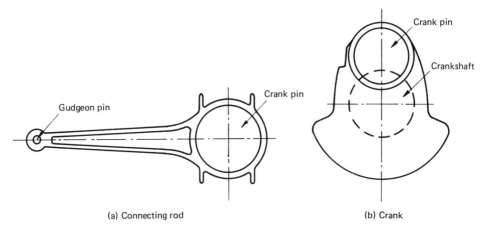

(a) Connecting rod   (b) Crank

Figure 7.2

7.2 Out-of-balance frame forces and moments

Let us now position the mechanism of Figure 7.1 within a general space frame as shown in Figure 7.3, such that the centreline, i.e. line of action of $\mathbf{FF}_n$, is inclined at some angle η to the vertical OY axis, and is positioned at some distance, z, from the origin O along the axis OZ. Also, for the general case, we shall denote the angle by

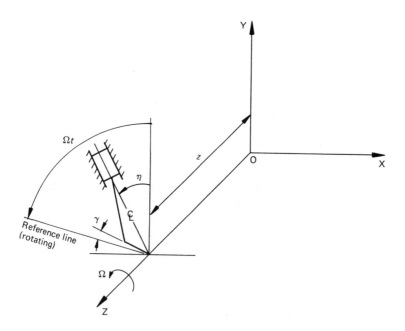

Figure 7.3

which the crank *lags* a rotating reference line as γ. We will neglect the non-time-dependent gravitational force. Therefore, in this case,

$$\mathbf{FF}_n = \Omega^2 C_n(-\mathbf{i}\sin\eta + \mathbf{j}\cos\eta)\cos n(\Omega t - \gamma - \eta)$$

which can be expressed in complex notation form as

$$\mathbf{FF}_n = \Omega^2 C_n \vec{\mathbf{F}}_n \cos n\Omega t \tag{7.10}$$

where

$$\vec{\mathbf{F}}_n = (-\mathbf{i}\sin\eta + \mathbf{j}\cos\eta)\underline{/-n(\gamma+\eta)}$$

and $\mathbf{i}$ and $\mathbf{j}$ are unit vectors along OX and OY respectively. Similarly, the moment vector of the nth order frame force about O, i.e. $\mathbf{MM}_n$, will be

$$\begin{aligned} \mathbf{MM}_n &= \mathbf{k}z \wedge \Omega^2 C_n \vec{\mathbf{F}}_n \cos n\Omega t \\ &= \Omega^2 C_n \vec{\mathbf{M}}_n \cos n\Omega t \end{aligned} \tag{7.11}$$

where

$$\vec{\mathbf{M}}_n = -z(\mathbf{i}\cos\eta + \mathbf{j}\sin\eta)\underline{/-n(\gamma+\eta)}$$

where $\mathbf{k}$ is unit vector along the axis OZ, and the sign $\wedge$ represents vector cross multiplication. Equations 7.10 and 7.11 describe the nth order frame force and frame force moment vectors produced by a single reciprocating mechanism. Let us now consider a system containing, say, m similar mechanisms driven by the same crankshaft and having the same value of C_n for each value of n; but having their own individual values of γ, η and z. For such a case equations 7.10 and 7.11 can be expanded to give the frame force and frame force moment vectors for the complete system as

$$\begin{aligned} \frac{\mathbf{FF}_n}{\Omega^2 C_n} &= \mathbf{i}\left[\sum_{k=1}^{m}(-\sin\eta_k)\underline{/-n(\gamma_k+\eta_k)}\right]\cos n\Omega t \\ &+ \mathbf{j}\left[\sum_{k=1}^{m}(\cos\eta_k)\underline{/-n(\gamma_k+\eta_k)}\right]\cos n\Omega t \end{aligned} \tag{7.12a}$$

$$\begin{aligned} \frac{\mathbf{MM}_n}{\Omega^2 C_n} &= \mathbf{i}\left[\sum_{k=1}^{m}(-z_k\cos\eta_k)\underline{/-n(\gamma_k+\eta_k)}\right]\cos n\Omega t \\ &+ \mathbf{j}\left[\sum_{k=1}^{m}(-z_k\sin\eta_k)\underline{/-n(\gamma_k+\eta_k)}\right]\cos n\Omega t \end{aligned} \tag{7.12b}$$

We will now proceed to apply the general equations 7.12a and 7.12b to investigate the nature of the unbalanced forces and moments in various types of positive displacement engines.

7.3   Regular in-line vertical cylinder engines

In such engines, the angle η as described in Section 7.2 is zero for all piston mechanisms. The cranks are displaced relative to each other by some fixed angle and it is convenient to set the value of γ for the first crank to zero. Furthermore we will assume the normal situation in which the axial distance, p, separating any adjacent pair of piston mechanisms is constant.

7.3.1 Four-cylinder engine (see Figure 7.4a)

This engine is arranged such that the cranks at planes 1 and 4 are set parallel and the cranks at planes 2 and 3 are set at 180° with reference to the cranks at planes 1 and 4 as shown in Figure 7.4b.

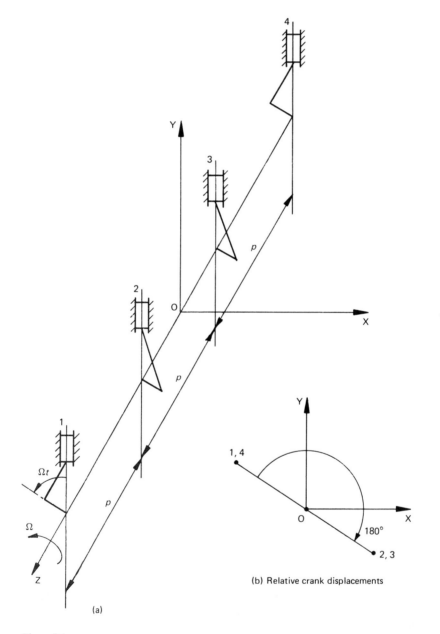

(b) Relative crank displacements

(a)

Figure 7.4

Therefore, the total nth order frame force vector $\mathbf{FF}_n$ is, from equation 7.12a,

$$\frac{\mathbf{FF}_n}{\Omega^2 C_n} = \mathbf{j}(\underline{/0^\circ} + \underline{/-n \cdot 180^\circ} + \underline{/-n \cdot 180^\circ} + \underline{/0^\circ})\cos n\Omega t$$

$$= 2\mathbf{j}(1 + \underline{/-n \cdot 180^\circ})\cos n\Omega t \tag{7.13}$$

From equation 7.13, the reader will deduce that the only frame force which is zero is the first order, i.e. $n = 1$. All higher order frame forces, i.e. $n = 2, 4$ and 6, are present. Equation 7.13 can therefore be written in the form

$$\frac{\mathbf{FF}_n}{\Omega^2 C_n} = 4\mathbf{j}\cos n\Omega t \qquad n = 2, 4, 6 \text{ etc.} \tag{7.14}$$

The total nth order frame force moment vector about O, $\mathbf{MM}_n$ is, from equation 7.12b,

$$\frac{\mathbf{MM}_n}{\Omega^2 C_n} = \mathbf{i}(-1.5p\underline{/0^\circ} - 0.5p\underline{/-n \cdot 180^\circ} + 0.5p\underline{/-n \cdot 180^\circ} + 1.5p\underline{/0^\circ})\cos n\Omega t$$

$$= 0 \text{ for all values of } n \tag{7.15}$$

For this configuration, therefore, there are *no frame force moments* about the axial mid-point O.

7.3.2 Five-cylinder engine (see Figure 7.5a)

With reference to the crank at plane 1, the cranks at planes 2, 3, 4 and 5 are angularly displaced by 144°, 72°, 216° and 288° respectively as shown in Figure 7.5b. Thus, from equation 7.12a, we have

$$\frac{\mathbf{FF}_n}{\Omega^2 C_n} = \mathbf{j}(\underline{/0^\circ} + \underline{/-n \cdot 144^\circ} + \underline{/-n \cdot 72^\circ} + \underline{/-n \cdot 216^\circ} + \underline{/-n \cdot 288^\circ})\cos n\Omega t \tag{7.16}$$

and upon expanding we find that

$$\frac{\mathbf{FF}_n}{\Omega^2 C_n} = 0 \qquad \text{for } n = 1, 2, 4, 6 \text{ etc.}$$

Similarly, from equation 7.12b, we have for the moments about O

$$\frac{\mathbf{MM}_n}{\Omega^2 C_n} = \mathbf{i}(-2p\underline{/0^\circ} - p\underline{/-n \cdot 144^\circ} + p\underline{/-n \cdot 216^\circ} + 2p\underline{/-n \cdot 288^\circ})\cos n\Omega t \tag{7.17}$$

and upon expanding we find that

$$\frac{\mathbf{MM}_1}{\Omega^2 p C_1} = \mathbf{i}3.374\underline{/-245.8^\circ}\cos \Omega t$$

$$\frac{\mathbf{MM}_2}{\Omega^2 p C_2} = \mathbf{i}3.69\underline{/-168.6^\circ}\cos 2\Omega t$$

$$\frac{\mathbf{MM}_4}{\Omega^2 p C_4} = \mathbf{i}3.374\underline{/-114.13^\circ}\cos 4\Omega t \tag{7.18}$$

$$\frac{\mathbf{MM}_6}{\Omega^2 p C_6} = \mathbf{i}3.374\underline{/-245.8^\circ}\cos 6\Omega t$$

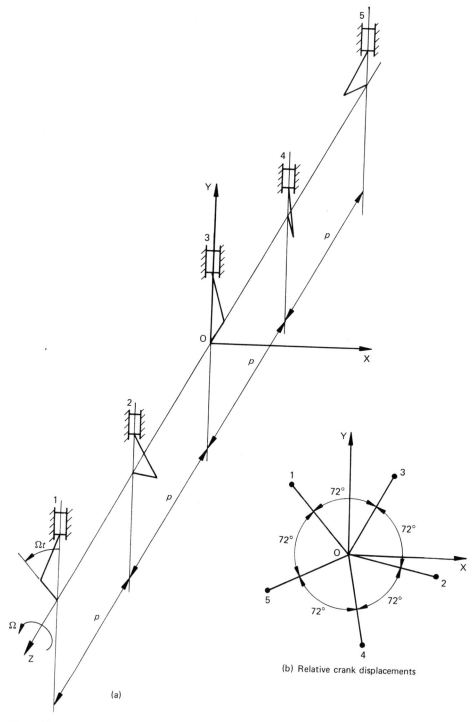

(a)

(b) Relative crank displacements

Figure 7.5

7.3.3 Six-cylinder engine (see Figure 7.6a)

In this engine arrangement the cranks at planes 2 and 5 are displaced by 240°, and the cranks at planes 3 and 4 are displaced by 120°, with respect to the parallel cranks at planes 1 and 6 as shown in Figure 7.6b.

From equation 7.12a we have, for this arrangement,

$$\frac{\mathbf{FF}_n}{\Omega^2 C_n} = 2\mathbf{j}(1 + \underline{/-n\cdot240°} + \underline{/-n\cdot120°})\cos n\Omega t \tag{7.19}$$

and expanding it is found that

$$\frac{\mathbf{FF}_1}{\Omega^2 C_1} = 0 \qquad \frac{\mathbf{FF}_2}{\Omega^2 C_2} = 0 \qquad \frac{\mathbf{FF}_4}{\Omega^2 C_4} = 0 \qquad \frac{\mathbf{FF}_6}{\Omega^2 C_6} = \mathbf{j}6\cos 6\Omega t \tag{7.20}$$

i.e. only the sixth order frame force exists. Considering now the frame force moment vector, $\mathbf{MM}_n$, about the axial mid-point, O, it can be shown by applying equation 7.12b that

$$\frac{\mathbf{MM}_n}{\Omega^2 pC_n} = 0 \quad \text{for } n = 1, 2, 4, 6 \text{ etc.} \tag{7.21}$$

From the results of equations 7.20 and 7.21 it can be seen that this arrangement represents a very smooth running engine.

7.4   Regular off-line cylinder engines

These engine configurations comprise two or more identical piston mechanisms, axially displaced along a common crankshaft. The line of action of the reciprocation is, however, generally not along the vertical axis, i.e. the angle η as described in Figure 7.3 is *not zero*.

Basically there are two forms by which such engines are arranged, namely those where each crank drives a single piston and those where any one crank drives two (or more) pistons. In the latter case when each crank drives two pistons, such engines are termed 'V engines' and the specific case of the V8 engine will be investigated in Section 7.4.2.

Let us for the moment start by investigating an off-line cylinder engine where each crank drives a single piston. One very common, and interesting, example of this is the 'flat-four boxer' engine.

7.4.1   The flat-four boxer engine (see Figure 7.7a)

This engine has four cylinders with the angle η equal to $\pm 90°$ as shown. The crank angles are displaced in a manner similar to the normal four cylinder in-line vertical engine described in Section 7.3.1, see Figure 7.7b.

Hence, from equation 7.12a, the nth order frame force vector, $\mathbf{FF}_n$, is

$$\frac{\mathbf{FF}_n}{\Omega^2 C_n} = \mathbf{i}[(-\sin 90°)\underline{/-n\cdot90°} + (\sin 90°)\underline{/-n\cdot(180° - 90°)}$$

$$+ (\sin 90°)\underline{/-n\cdot(180° - 90°)} + (-\sin 90°)\underline{/-n\cdot90°}]\cos n\Omega t \tag{7.22}$$

$$= 0 \qquad \text{for all values of } n \tag{7.23}$$

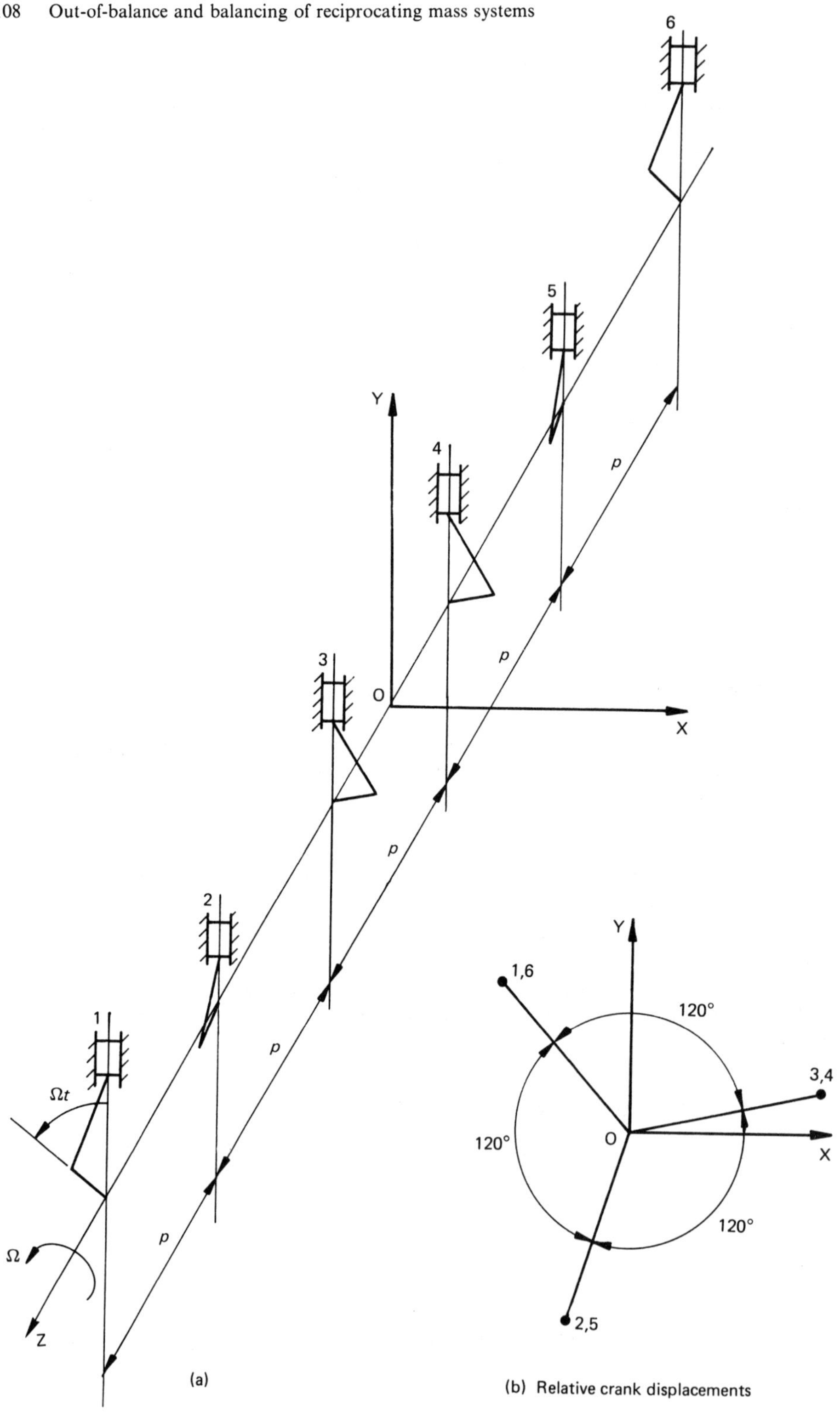

(a)

(b) Relative crank displacements

Figure 7.6

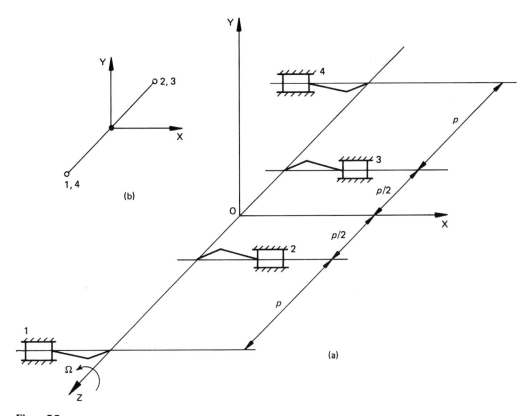

Figure 7.7

Similarly, by applying equation 7.12b, the nth order frame force moment vector about O, $\mathbf{MM}_n$, is

$$\frac{\mathbf{MM}_n}{\Omega^2 C_n} = \mathbf{j}[(-1.5p\sin 90°)\underline{/-n\cdot 90°} + (0.5p\sin 90°)\underline{/-n\cdot(180°-90°)}$$

$$+ (-0.5p\sin 90°)\underline{/-n\cdot(180°-90°)} + (1.5p\sin 90°\underline{/-n\cdot 90°})]\cos n\Omega t$$

$$= 0 \quad \text{for all values of } n \tag{7.24}$$

i.e. this engine represents a state of 'natural balance', which is the commercial claim made by the manufacturers of such an engine, e.g. Alfa Romeo.

7.4.2 V8 engine

Figure 7.8a shows a typical V8 engine arrangement where each of the four cranks drives two pistons simultaneously and the 'V angle', 2η, in this case will be taken to be 90°. Relative to the crank driving pistons 1 and 2, the cranks driving pistons 3 and 4, 5 and 6, and 7 and 8 are angularly displaced by 270°, 90° and 180° respectively as shown in Figure 7.8b. Thus applying equation 7.12a to this arrangement (with the data listed in Table 7.1) gives

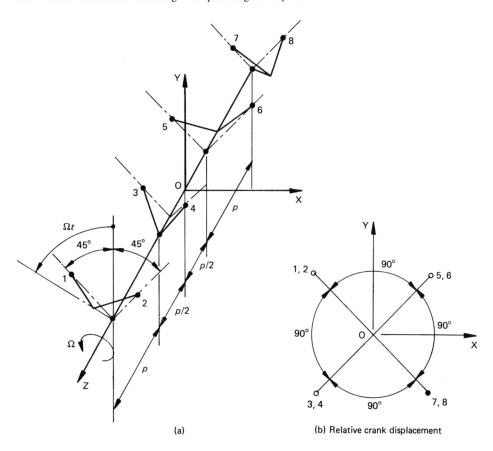

(a)   (b) Relative crank displacement

Figure 7.8

Table 7.1

	Piston							
	1	2	3	4	5	6	7	8
γ	0	0	270	270	90	90	180	180
η	45	−45	45	−45	45	−45	45	−45
z	$1.5p$	$1.5p$	$0.5p$	$0.5p$	$−0.5p$	$−0.5p$	$−1.5p$	$−1.5p$

$$\frac{FF_n}{\Omega^2 C_n} = i[(-\sin 45° \underline{/-n\cdot 45°}) + (\sin 45° \underline{/n\cdot 45°}) + (-\sin 45° \underline{/-n\cdot 315°})$$

$$+ (\sin 45° \underline{/-n\cdot 225°}) + (-\sin 45° \underline{/-n\cdot 135°})$$

$$+ (\sin 45° \underline{/-n\cdot 45°}) \ + (-\sin 45° \underline{/-n\cdot 225°})$$

$$+ (\sin 45° \underline{/-n\cdot 135°})] \cos n\Omega t \qquad\qquad (7.25)$$

$$+ j[(\cos 45° \underline{/-n\cdot 45°}) + (\cos 45° \underline{/n\cdot 45°}) + (\cos 45° \underline{/-n\cdot 315°})$$

$$+ (\cos 45° \underline{/-n\cdot 225°}) + (\cos 45° \underline{/-n\cdot 135°}) + (\cos 45° \underline{/n\cdot 45°})$$

$$+ (\cos 45° \underline{/-n\cdot 225°}) \ + (\cos 45° \underline{/-n\cdot 135°})] \cos n\Omega t$$

Expanding equation 7.25 we find that, for values of $n = 1, 2, 4$ and 6,

$$\frac{FF_1}{\Omega^2 C_1} = 0 \qquad\qquad \frac{FF_2}{\Omega^2 C_2} = 0$$

$$\frac{FF_4}{\Omega^2 C_4} = j\, 5.656 \underline{/-180^\circ} \cos 4\Omega t \qquad \frac{FF_6}{\Omega^2 C_6} = 0 \qquad\qquad (7.26)$$

Consider now the frame force moment vector, MM_n, about the axial centre point O, which from equation 7.12b is

$$\begin{aligned}
\frac{MM_n}{\Omega^2 p C_n} = i[&-1.5(\cos 45^\circ \underline{/-n\cdot 45^\circ}) - 1.5(\cos 45^\circ \underline{/n\cdot 45^\circ})\\
&- 0.5(\cos 45^\circ \underline{/-n\cdot 135^\circ}) - 0.5(\cos 45^\circ \underline{/-n\cdot 225^\circ})\\
&+ 0.5(\cos 45^\circ \underline{/-n\cdot 135^\circ}) + 0.5(\cos 45^\circ \underline{/-n\cdot 45^\circ})\\
&+ 1.5(\cos 45^\circ \underline{/-n\cdot 225^\circ}) + 1.5(\cos 45^\circ \underline{/-n\cdot 135^\circ})]\cos n\Omega t \qquad (7.27)\\
+ j[&-1.5(\sin 45^\circ \underline{/-n\cdot 45^\circ}) - 1.5(-\sin 45^\circ \underline{/n\cdot 45^\circ})\\
&- 0.5(\sin 45^\circ \underline{/-n\cdot 315^\circ}) - 0.5(-\sin 45^\circ \underline{/-n\cdot 225^\circ})\\
&+ 0.5(\sin 45^\circ \underline{/-n\cdot 135^\circ}) + 0.5(-\sin 45^\circ \underline{/-n\cdot 45^\circ})\\
&+ 1.5(\sin 45^\circ \underline{/-n\cdot 225^\circ}) + 1.5(-\sin 45^\circ \underline{/-n\cdot 135^\circ})]\cos n\Omega t
\end{aligned}$$

from which it is calculated that

$$\begin{aligned}
\frac{MM_1}{\Omega^2 p C_1} &= 3.162(i \underline{/-161.5^\circ} + j \underline{/-251.5^\circ}) \cos \Omega t\\
&= 3.162 \underline{/-251.5^\circ}(-i\sin \Omega t + j\cos \Omega t) \qquad\qquad (7.28)
\end{aligned}$$

and

$$\frac{MM_2}{\Omega^2 p C_2} = \frac{MM_4}{\Omega^2 p C_4} = \frac{MM_6}{\Omega^2 p C_6} = 0$$

In conclusion, from equations 7.26 and 7.28, this engine exhibits only a fourth-order frame force (which is usually insignificant) and a primary frame force moment which can be represented diagrammatically as shown in Figure 7.9. As we will see in Section 7.5, this primary frame force moment can be easily balanced.

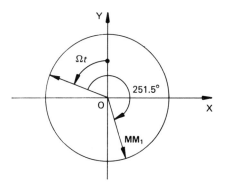

Figure 7.9

7.5 Balancing of primary frame force and frame force moment

So far we have investigated the nature of the frame forces and frame force moments present in various engine arrangements. Prior to investigating methods of balancing such forces and moments, let us first consider the frame force and frame force moment vectors, **BF** and **BM** respectively, arising from placing two equal rotating masses, value m, offset from the crankshaft axis by e, and located at either end of the crankshaft of a general multi-cylinder engine as shown in Figure 7.10 where, as before, the reference line indicated corresponds to the instantaneous angular position of some reference crank, usually that driving piston number 1.

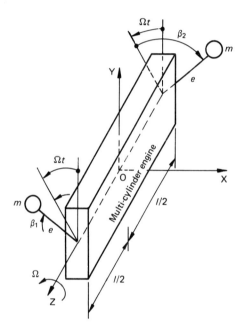

Figure 7.10

From equations 6.3a the frame force vector **BF** is in this case

$$\frac{\mathbf{BF}}{\Omega^2} = me(\underline{/+\beta_1} + \underline{/-\beta_2})(-\mathbf{i}\sin\Omega t + \mathbf{j}\cos\Omega t) \tag{7.29}$$

Similarly, from equation 6.3b, the frame force moment vector, **BM**, about the axial centre of the engine, O, is

$$\frac{\mathbf{BM}}{\Omega^2} = \frac{mel}{2}(\underline{/-\beta_2} - \underline{/+\beta_1})(\mathbf{i}\cos\Omega t + \mathbf{j}\sin\Omega t) \tag{7.30}$$

Now, if $\beta_2 = -\beta_1$, equation 7.29 becomes

$$\frac{\mathbf{BF}}{\Omega^2} = 2me\underline{/+\beta_1}(-\mathbf{i}\sin\Omega t + \mathbf{j}\cos\Omega t) \tag{7.31}$$

and equation 7.30 becomes

$$\frac{\mathbf{BM}}{\Omega^2} = 0 \tag{7.32}$$

Therefore if the multi-cylinder engine alone (in the absence of these rotating masses) was found to produce an unacceptably high *primary frame force*, this could be reduced by superimposing upon the system two equal rotating masses arranged as shown in Figure 7.11, with suitably selected values of m, e and β_1. Furthermore, in so doing,

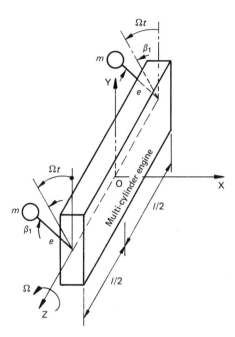

Figure 7.11

we do not introduce any additional frame force moment. Now, if $\beta_2 = \pi - \beta_1$, equations 7.29 and 7.30 reduce to

$$\mathbf{BF} = 0 \tag{7.33}$$

and

$$\frac{\mathbf{BM}}{\Omega^2} = -mel\underline{/+\beta_1}(\mathbf{i}\cos\Omega t + \mathbf{j}\sin\Omega t) \tag{7.34}$$

Thus, in this case, if the multi-cylinder engine alone was found to produce an unacceptably high primary frame force moment, this could be reduced by super-imposing upon the system two equal rotating masses arranged as shown in Figure 7.12, with suitably selected values of m, e and β_1.

Furthermore, this added arrangement does not give rise to any additional frame forces. The following numerical example illustrates the above primary balancing procedure.

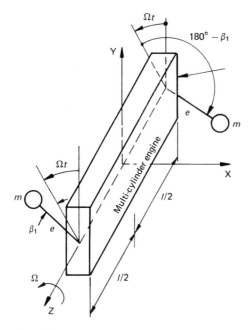

Figure 7.12

Example 7.1

A large vertical five-cylinder marine diesel engine has total effective reciprocating masses at each piston of 432 kg and the relative crank angles are described in Figure 7.5b. The cranks are 305 mm long with equal axial spacing of 610 mm.

Calculate the mass and angular position, relative to the crank at plane 1, of balance bobs which should be placed at planes 1 and 5 in order that the amplitude of the inherent primary frame force moment about the axis OX be reduced by 40 per cent, assuming that the bobs are placed at 305 mm offset from the crankshaft rotational axis.

SOLUTION

For the engine alone, equation 7.18 gives the primary frame force moment vector, $\mathbf{MM}_1$, about O as

$$\frac{\mathbf{MM}_1}{\Omega^2} = \mathbf{i}\, 3.374\, p C_1 \underline{/-254.8°}\cos\Omega t$$

where $p = 0.61$ m; and from equation 7.9

$$C_1 = 432 \times 0.305 = 131.76\,\text{kg m}$$

giving

$$\frac{\mathbf{MM}_1}{\Omega^2} = \mathbf{i}\, 271.18 \underline{/-254.8°}\cos\Omega t\,\text{kg m}^2 \qquad (7.35)$$

In order to reduce the amplitude of the primary frame force moment (about the

OX axis) we will superimpose upon the system the rotating offset mass system described in Figure 7.12, and only concern ourselves with the **i** component of the corresponding frame force moment vector **BM** described in equation 7.34. Hence summing equation 7.35 and this truncated form of equation 7.34, the resultant frame force moment vector, $\mathbf{RM}_1$, becomes

$$\frac{\mathbf{RM}_1}{\Omega^2} = \mathbf{i}\,271.18\underline{/-245.8°}\cos\Omega t - \mathbf{i}mel\underline{/+\beta_1}\cos\Omega t \; \text{kg\,m} \tag{7.36}$$

and it is required that

$$\frac{\mathbf{RM}_1}{\Omega^2} = \frac{0.6 \times \mathbf{MM}_1}{\Omega^2} = \mathbf{i}(0.6 \times 271.18)\underline{/-245.8°}\cos\Omega t \; \text{kg\,m}^2 \tag{7.37}$$

Therefore equating equations 7.37 and 7.36 gives

$$\beta_1 = -245.8°$$

and

$$m = \frac{0.4 \times 271.18}{el}$$

$$= \frac{0.4 \times 271.18}{0.305 \times 2.44}$$

$$= 145.8 \, \text{kg}$$

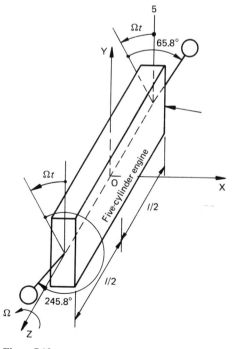

Figure 7.13

Hence at plane 1 we must place a 305 mm offset mass of 145.8 kg at 245.8° clockwise to the crank at plane 1, i.e. in the opposite direction to that of rotation. At plane 5, therefore, we place a similar mass, at the same offset, at 65.8° (245.8° − 180°) clockwise from the crank at plane 1, or 137.8° clockwise from the crank at plane 5 as shown in Figure 7.13.

If we now re-examine equation 7.34 describing the frame force moment vector, **BM**, we notice that whereas the **i** component has been used to reduce the amplitude of **MM**$_1$, we have now introduced an additional frame force moment component, namely the **j** component of equation 7.34. This will result in the introduction of a moment of amplitude $108.5\Omega^2$ N m acting about the vertical OY axis, which prior to this balancing exercise did not exist. Therefore we could claim that in effect we are *not* reducing the degree of out-of-balance, but rather altering the form of it.

Problems

7.1   Figure 7.14 shows a vertical twin-cylinder engine. The pistons are connected to a 'two-throw' crankshaft, the cranks being 180° to each other.

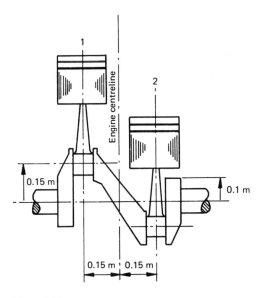

Figure 7.14

Develop expressions for the out-of-balance forces and couples produced by the reciprocating effects.

Such an engine has the following details:

- Reciprocating mass/cylinder = 2.00 kg.
- Crank and balance weight radius = 0.15 m.
- Con-rod length = 0.35 m.
- Pitch of cylinders from centreline = 0.15 m.
- Harmonic coefficients: $A_1 = 1$; $A_2 = +0.428$; $A_4 = -0.020$.

Calculate the values of the significant out-of-balance effects up to and including the fourth order, at an engine speed of 1200 rev/min. Also calculate the mass and position of balance masses in planes 0.25 m on each side of the engine centreline, for equal vertical and horizontal balance.

Answer: $\mathbf{MM}_1 = \mathbf{i}\,1421\,\underline{/180°}\,\text{N m}; \mathbf{FF}_2 = \mathbf{j}\,4055\,\text{N}; \mathbf{FF}_4 = \mathbf{j}\,189.5\,\underline{/180°}\,\text{N}; 0.6\,\text{kg}$ set at 180° from the adjacent cranks.

7.2   A six-cylinder V-type engine has the banks of cylinders arranged at $\pm 30°$ about a vertical centreline.

- The cranks are at 120°.
- The crank sequence is 1, 2, 3.
- Axial spacing = 300 mm.
- Crank radius = 50 mm.
- Reciprocating mass is 1.4 kg per cylinder.
- Engine speed = 2000 rev/min.
- Harmonic constants: $A_1 = 1$; $A_2 = 0.272$; $A_4 = -0.005$.

Calculate all out-of-balance frame forces up to and including the fourth order. Calculate also the first-order out-of-balance couple, and the position and magnitude of balance masses at crank radius in the two outer lines to give zero horizontal out-of-balance couple.

Answer: $\mathbf{FF}_1 = \mathbf{FF}_2 = \mathbf{FF}_4 = 0$; $\mathbf{MM}_1 = \mathbf{i}\,2.394\,\underline{/-210°} + \mathbf{j}\,0.8\,\underline{/-300°}\,\text{kN/m}$; plane 1 (pistons 1 and 2), 1.8 kg at 210° clockwise from crank at 1; plane 3 (pistons 5 and 6), 1.8 kg at 210° anticlockwise from crank at 3.

7.3   A regular six-cylinder two-stroke oil engine (cranks at 60°) can be constructed with the alternative cranks orders (a) 1 4 2 6 3 5 or (b) 1 5 3 4 2 6.

Investigate primary and secondary out-of-balance and indicate which crank order you would recommend, assuming that it is possible to add balance masses to the crankshaft. For your preferred arrangement, calculate the size and position of any balance masses you suggest.

- Reciprocating mass/cylinder = 250 kg.
- Crank radius = 300 mm.
- Axial pitch of cylinders = 600 mm.
- $A_2 = 0.27$.
- Balance masses at axial pitch of 3.5 m and 300 mm radius.

Answer: (a) $\mathbf{FF}_1 = \mathbf{FF}_2 = \mathbf{MM}_2 = 0$; $\mathbf{MM}_1 = \mathbf{i}\,156\,\Omega^2\,\underline{/-210°}\,\text{N}$; (b) $\mathbf{FF}_1 = \mathbf{FF}_2 = \mathbf{MM}_1 = 0$; $\mathbf{MM}_2 = \mathbf{i}\,42\,\Omega^2\,\underline{/-210°}\,\text{N}$.

8

Introduction to dynamics of general space motion

8.1 Introduction

In Chapter 3 attention was focused on the dynamics of bodies undergoing motion contained within a single plane, and the application of this study to coplanar mechanisms was demonstrated. In modern times, however, in the design of systems such as space satelites, aircraft and industrial robots, it is necessary to apply an understanding of the dynamics associated with bodies undergoing general space motion, i.e. time-dependent motion which has components along the mutually perpendicular axes OX, OY and OZ of the normal coordinate system.

Consider a particle, S, of some rigid body undergoing general space motion as shown in Figure 8.1, where h is the radial distance of the particle from the origin of the axis system, O, and ϕ and ψ are the Euler angles of *nutation* and *precession* respectively. Therefore the instantaneous displacement of S relative to O is the displacement vector, $\boldsymbol{\delta}_{\mathrm{OS}}$, such that

$$\boldsymbol{\delta}_{\mathrm{OS}} = h(\mathbf{i} \sin \phi \sin \psi + \mathbf{j} \cos \phi + \mathbf{k} \sin \phi \cos \psi) \tag{8.1}$$

where $\mathbf{i}, \mathbf{j}$ and $\mathbf{k}$ are unit vectors along OX, OY and OZ respectively.

Now, since the motion is time dependent, we will for the purpose of future analysis adopt the convention that

$$\frac{dh}{dt} = \dot{h} \qquad \frac{d\phi}{dt} = \dot{\phi} \qquad \frac{d\psi}{dt} = \dot{\psi}$$

$$\frac{d^2h}{dt^2} = \ddot{h} \qquad \frac{d^2\phi}{dt^2} = \ddot{\phi} \qquad \frac{d^2\psi}{dt^2} = \ddot{\psi}$$

Having described the displacement of the body particle from the origin, we can now proceed to investigate the kinematics and kinetics associated with the body particle.

8.2 Kinematic analysis

In Section 8.1, we described the instantaneous displacement vector, d_{OS}, of a particle, S, undergoing general space motion relative to the normal axes origin O, see Figure 8.1 and equation 8.1. On the basis of this displacement vector, we can now proceed to described the instantaneous velocity and acceleration vectors for the particle undergoing such motion.

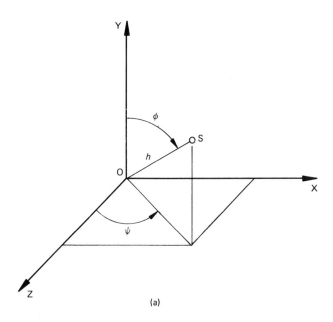

(a)

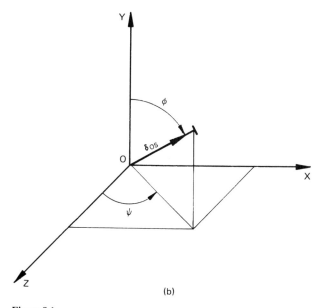

(b)

Figure 8.1

8.2.1 Velocity vectors

Considering equation 8.1 which describes the instantaneous displacement vector, the velocity of the particle point S relative to O can be written as the vector $\mathbf{v}_{OS}$ such that

$$
\begin{aligned}
\mathbf{v}_{OS} &= \frac{\mathrm{d}\boldsymbol{\delta}_{OS}}{\mathrm{d}t} \\
&= \dot{h}(\mathbf{i}\sin\phi\sin\psi + \mathbf{j}\cos\phi + \mathbf{k}\sin\phi\cos\psi) \\
&\quad + h\dot{\phi}(\mathbf{i}\cos\phi\sin\psi - \mathbf{j}\sin\phi + \mathbf{k}\cos\phi\cos\psi) \\
&\quad + h\dot{\psi}\sin\phi(\mathbf{i}\cos\psi - \mathbf{k}\sin\psi)
\end{aligned}
\tag{8.2}
$$

8.2.2 Acceleration vectors

Differentiating equation 8.2 describing the velocity vector of S relative to O, $\mathbf{v}_{OS}$, with respect to time and grouping like terms gives the acceleration vector of S relative to O, $\mathbf{a}_{OS}$, such that

$$
\begin{aligned}
\mathbf{a}_{OS} &= \mathrm{d}\mathbf{v}_{OS}/\mathrm{d}t \\
&= \ddot{h}(\mathbf{i}\sin\phi\sin\psi + \mathbf{j}\cos\phi + \mathbf{k}\sin\phi\cos\psi) \\
&\quad + 2\dot{h}\dot{\phi}(\mathbf{i}\cos\phi\sin\psi - \mathbf{j}\sin\phi + \mathbf{k}\cos\phi\cos\psi) + 2\dot{h}\dot{\psi}\sin\phi(\mathbf{i}\cos\psi - \mathbf{k}\sin\psi) \\
&\quad + h\ddot{\phi}(\mathbf{i}\cos\phi\sin\psi - \mathbf{j}\sin\phi + \mathbf{k}\cos\phi\cos\psi) + h\ddot{\psi}\sin\phi(\mathbf{i}\cos\psi - \mathbf{k}\sin\psi) \\
&\quad - h\dot{\phi}^2(\mathbf{i}\sin\phi\sin\psi + \mathbf{j}\cos\phi + \mathbf{k}\sin\phi\cos\psi) - h\dot{\psi}^2\sin\phi(\mathbf{i}\sin\psi + \mathbf{k}\cos\psi) \\
&\quad + 2h\dot{\phi}\dot{\psi}\cos\phi(\mathbf{i}\cos\psi - \mathbf{k}\sin\psi)
\end{aligned}
\tag{8.3}
$$

The reader may, at this stage, wish to compare the complexity of the velocity and acceleration vector equations, i.e. equations 8.2 and 8.3 respectively, with the corresponding equations pertaining to general plane motion, i.e. equations 3.3 and 3.4 of Chapter 3, although the various radial and tangential components of velocity and acceleration, and centripetal and Coriolis components of acceleration, are easily identifiable from inspection of equations 8.2 and 8.3. Suffice to say that for problems involving general space motion it is not convenient to construct velocity and acceleration vector diagrams as in the case of general plane motion. However, with the aid of modern computer facilities, problems involving both types of motion (general plane and general space) can be readily handled.

8.3 Kinetic analysis

In Section 8.2 we derived expressions for the velocity and acceleration vectors associated with a body particle undergoing general space motion (see equations 8.2 and 8.3). Let us now turn our attention to the forces and moments necessary to produce such acceleration, or changes in velocity.

8.3.1 Force and moments

Once again consider the body particle, S, undergoing general space motion as shown in Figure 8.1. If the mass of such a particle is δm, and given that equation 8.3 describes the acceleration vector of the particle with respect to the origin, O, namely $\mathbf{a}_{OS}$, then by Newton's Second Law of Motion the net force vector, **f**, required to produce this acceleration is

$$\mathbf{f} = \delta m \, \mathbf{a}_{OS} \tag{8.4}$$

Thus the net force vector, **F**, required to accelerate all such particles within the rigid body containing them will be

$$\mathbf{F} = \rho \int_{\text{vol}} \mathbf{a}_{OS} \, \delta v \tag{8.5}$$

where δv is the particle volume given by

$$\delta v = h^2 \delta\phi \, \delta\psi \, \delta h \tag{8.6}$$

and ρ is the mass density of the body material. Hence

$$\mathbf{F} = \rho \int\int\int \mathbf{a}_{OS} h^2 \, d\phi \, d\psi \, dh \tag{8.7}$$

Therefore the inertia force vector, **IF**, will be

$$\mathbf{IF} = -\mathbf{F} = -\rho \int\int\int \mathbf{a}_{OS} h^2 \, d\phi \, d\psi \, dh \tag{8.8}$$

Consider now the net moment vector, **m**, about the origin O required to produce this particle acceleration, which is

$$\mathbf{m} = \boldsymbol{\delta}_{OS} \wedge \delta m \, \mathbf{a}_{OS} \tag{8.9}$$

where the symbol $\wedge$ denotes vector cross multiplication. Thus, once again, the total net moment vector, **M**, about O required to accelerate the rigid body containing all such particles will be

$$\mathbf{M} = \rho \int\int\int \boldsymbol{\delta}_{OS} \wedge \mathbf{a}_{OS} h^2 \, d\phi \, d\psi \, dh \tag{8.10}$$

and the inertia moment (or torque) vector, **IT**, will be

$$\mathbf{IT} = -\mathbf{M} = -\rho \int\int\int \boldsymbol{\delta}_{OS} \wedge \mathbf{a}_{OS} h^2 \, d\phi \, d\psi \, dh \tag{8.11}$$

Note that the integrations are all performed over the geometric bounds of the body.

8.3.2   Momentum and moment of momentum

Let us reconsider equation 8.4 which describes the application of Newton's Second Law of Motion to the particle of mass δm. More generally, this equation can be written as

$$\mathbf{f} = \frac{d(\delta m \, \mathbf{v}_{OS})}{dt} \tag{8.12}$$

i.e. the net force producing the particle acceleration can be redescribed as the *rate of change of momentum* of the particle. Subsequently, the net body force and inertia force vectors, **F** and **IF**, respectively, can be obtained by integrating equation 8.12 over the total volume of the body.

Consider now the *rate of change of moment of momentum* of the particle about the origin O, i.e.

$$\frac{d(\delta \mathbf{H})}{dt} = \frac{d}{dt}(\boldsymbol{\delta}_{OS} \wedge \delta m \, \mathbf{v}_{OS}) \tag{8.13}$$

where $\delta \mathbf{H}$ is the moment of momentum, often referred to as the angular momentum of the particle about O.

Performing the differentiation we have

$$\frac{d(\delta \mathbf{H})}{dt} = \left(\frac{d}{dt}\boldsymbol{\delta}_{OS} \wedge \delta m \, \mathbf{v}_{OS}\right) + \left(\boldsymbol{\delta}_{OS} \wedge \frac{d}{dt}\delta m \, \mathbf{v}_{OS}\right) \tag{8.14}$$

Now, since $d\boldsymbol{\delta}_{OS}/dt = \mathbf{v}_{OS}$, the first term on the right-hand side of equation 8.14 is zero, hence

$$\frac{d(\delta \mathbf{H})}{dt} = \left(\boldsymbol{\delta}_{OS} \wedge \frac{d}{dt}\delta m \, \mathbf{v}_{OS}\right) = \frac{d}{dt}\left(\boldsymbol{\delta}_{OS} \wedge \delta m \, \mathbf{v}_{OS}\right) \tag{8.15}$$

If we now compare equations 8.15, 8.9 and 8.12, we deduce that

$$\mathbf{m} = \frac{d(\delta \mathbf{H})}{dt} \tag{8.16}$$

Subsequently, by integration of all such particles contained within the body, we have devised an alternative method (and a very convenient one) of obtaining an expression for the net moment of force about O required to accelerate the body.

Although, as mentioned in Section 8.1, there are numerous modern applications involving the study of bodies undergoing general space motion, perhaps one of the most common, and technically fascinating, of all applications is that of the **gyroscope** and the **gyroscopic torque** to which it inevitably gives rise. Furthermore, in our forthcoming study of such systems, extensive use will be made of the relationship described by equation 8.16. In view of this, and prior to a detailed study of gyroscopes and gyroscopic torques, let us derive expressions for the vectors describing the moment of momentum of two simple systems.

8.3.3 Solid uniform disc spinning about its central axis

Figure 8.2 illustrates the case in question where a solid disc of radius R and thickness b is spinning at a constant rate of Ω_s rad/s, in the sense indicated, about it own central polar axis OZ'. For the sake of future reference let us denote the normal axes OX', OY' and OZ' as the local axes system of the disc and let $\mathbf{i}'$, $\mathbf{j}'$ and $\mathbf{k}'$ be unit vectors along these axes respectively. Similarly, OX, OY and OZ are the fixed reference axes and $\mathbf{i}$, $\mathbf{j}$ and $\mathbf{k}$ are unit vector along these axes. Furthermore, since the angles ϕ' and ψ' shown describe the orientation between the two sets of axes and assuming that the axis OX' is maintained parallel to the plane enclosed by the reference axes OX and OZ, then

$$\begin{aligned}
\mathbf{i}' &= \mathbf{i}\cos\psi' - \mathbf{k}\sin\psi' \\
\mathbf{j}' &= -\mathbf{i}\cos\phi'\sin\psi' + \mathbf{j}\sin\phi' - \mathbf{k}\cos\phi'\cos\psi' \\
\mathbf{k}' &= \mathbf{i}\sin\phi'\sin\psi' + \mathbf{j}\cos\phi' + \mathbf{k}\sin\phi'\cos\psi'
\end{aligned} \tag{8.17}$$

Consider now the displacement vector, $\boldsymbol{\delta}_{OS}$, of the particle S, i.e.

$$\boldsymbol{\delta}_{OS} = \mathbf{i}'r\cos\theta + \mathbf{j}'r\sin\theta \tag{8.18}$$

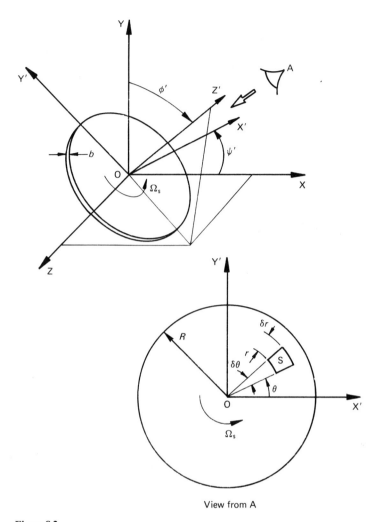

View from A

Figure 8.2

and, since $\theta = \Omega_s t$, the corresponding velocity vector, $\mathbf{v}_{os}$, is

$$\mathbf{v}_{os} = \frac{\mathrm{d}(\boldsymbol{\delta}_{os})}{\mathrm{d}t} = -\mathbf{i}'r\Omega_s \sin\theta + \mathbf{j}'r\Omega_s \cos\theta \tag{8.19}$$

Now, since the mass δm of the particle is $\rho br\,\delta r\,\delta\theta$, equations 8.18 and 8.19 can be multiplied (vectorially) to produce an expression for the particle moment of momentum, $\delta\mathbf{H}$, i.e.

$$\delta\mathbf{H} = \boldsymbol{\delta}_{os} \wedge \rho br\,\delta r\,\delta\theta\,\mathbf{v}_{os} = \mathbf{k}'\rho br^3\Omega_s\,\delta r\,\delta\theta \tag{8.20}$$

Therefore the total moment of momentum, $\mathbf{H}$, of the solid disc is

$$\mathbf{H} = \mathbf{k}'\rho b\Omega_s \int_0^{2\pi}\int_0^R r^3\,\mathrm{d}r\,\mathrm{d}\theta = \frac{\mathbf{k}'\rho b\pi R^4}{2}\Omega_s \tag{8.21}$$

noting that $\rho b\pi R^4/2$ is the polar mass moment of inertia of the disc (about $O'Z$) $= I_0$.

Equation 8.21 implies that the moment of momentum of such a spinning disc can be represented as a vector of magnitude equal to its angular momentum acting along the polar axis of the body and in a direction in accordance with the 'right-hand corkscrew rule' (if the present axis system is adopted) as shown in Figure 8.3.

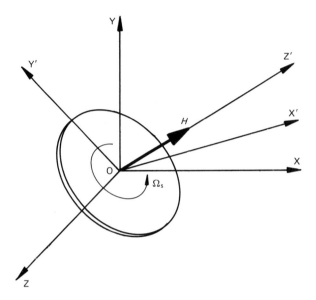

Figure 8.3

It follows therefore, by substituting for the unit vector $\mathbf{k}'$ from equation 8.17, that we can re-express the moment of momentum vector in terms of unit vectors along the fixed reference axes, i.e.

$$\mathbf{H} = I_0\Omega_s(\mathbf{i}\sin\phi'\sin\psi' + \mathbf{j}\cos\phi' + \mathbf{k}\sin\phi'\cos\psi') \tag{8.22}$$

8.3.4 Solid uniform spinning disc undergoing uniform precession at constant nutation

Consider now the solid spinning disc described in Figure 8.2, moving about the fixed axis system as shown in Figure 8.4, such that the rate of change of the angle of precession, $\dot\psi$, is constant and the angle of nutation, ϕ, is maintained at a constant value.

Now the displacement vector, $\boldsymbol{\delta}_{OS}$, which describes the displacement of the particle S relative to the fixed reference axes centre O, is

$$\boldsymbol{\delta}_{OS} = \boldsymbol{\delta}_{OO'} + \boldsymbol{\delta}_{O'S}$$

Substituting from equations 8.1 and 8.18 we have

$$\boldsymbol{\delta}_{OS} = h(\mathbf{i}\sin\phi\sin\psi + \mathbf{j}\cos\phi + \mathbf{k}\sin\phi\cos\psi) + \mathbf{i}'r\cos\theta + \mathbf{j}'r\sin\theta \tag{8.23}$$

where once again $\mathbf{i}'$ and $\mathbf{j}'$ are unit vectors along the local axes O'X' and O'Y'

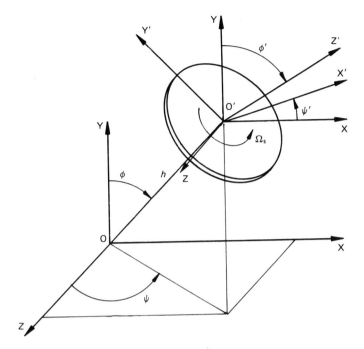

Figure 8.4

respectively. Now if $\dot{\psi} = \Omega_p$, and h is constant, then

$$\mathbf{v}_{OS} = h(\mathbf{i}\Omega_p \sin\phi \cos\psi - \mathbf{k}\Omega_p \sin\phi \sin\psi) + \frac{d}{dt}(\mathbf{i}'r\cos\theta + \mathbf{j}'r\sin\theta) \tag{8.24}$$

and, since, as before, the particle mass is $\rho br\,\delta r\,\delta\theta$, the particle moment of momentum, $\delta\mathbf{H}$, is

$$\delta\mathbf{H} = \boldsymbol{\delta}_{OS} \wedge \rho br\,\delta r\,\delta\theta\,\mathbf{v}_{OS} \tag{8.25}$$

Therefore the moment of momentum of the disc, $\mathbf{H}$, becomes

$$\mathbf{H} = \rho b \int_0^{2\pi} \int_0^R \boldsymbol{\delta}_{OS} \wedge r\mathbf{v}_{OS}\,dr\,d\theta \tag{8.26}$$

Substituting for $\mathbf{i}', \mathbf{j}'$ and $\mathbf{k}'$ from equation 8.17, equations 8.23–8.26 could be solved for general values of ϕ, ψ, ϕ' and ψ'.

Alternatively, the moment of momentum of the disc may be obtained by calculating the component of the moment of momentum of the disc associated only with the uniform precession at Ω_p, i.e. in the absence of its own central axis spin at Ω_s and subsequently adding to this the component of the moment of momentum of the disc associated with its own central spin only. This latter component may be obtained by substituting the relevant values of ψ' and ϕ' in equation 8.22. Let us now apply this procedure in order to derive expressions for the moment of momentum for three specific cases.

(a) $\phi' = \phi$ and $\psi' = \psi$. In this case the line OO' and axis O'Z' are parallel, i.e. OO' is perpendicular to the face of the disc as shown in Figure 8.5. Let us commence by deriving an expression for the component of the moment of momentum of the body, $\mathbf{H}_1$, associated with the uniform precession only, i.e. set $\Omega_s = 0$.

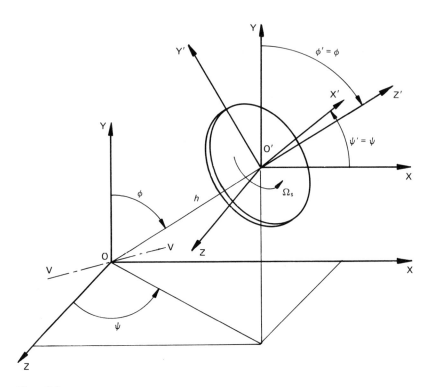

Figure 8.5

Hence substituting in equations 8.17, 8.23 and 8.24, we have

$$\boldsymbol{\delta}_{os} = \mathbf{i}(h \sin \phi \sin \psi - r \sin \theta \cos \phi \sin \psi + r \cos \theta \cos \psi) + \mathbf{j}(h \cos \phi + r \sin \theta \sin \phi)$$
$$+ \mathbf{k}(h \sin \phi \cos \psi - r \sin \theta \cos \phi \cos \psi - r \cos \theta \sin \psi) \tag{8.27}$$

and

$$\mathbf{v}_{os} = \mathbf{i}(h\Omega_p \sin \phi \cos \psi - r\Omega_p \sin \theta \cos \phi \cos \psi - r\Omega_p \cos \theta \sin \psi)$$
$$+ \mathbf{k}(-h\Omega_p \sin \phi \sin \psi + r\Omega_p \sin \theta \cos \phi \sin \psi - r\Omega_p \cos \theta \cos \psi) \tag{8.28}$$

remembering that $\dot{\psi} = \Omega_p$ and $\dot{\phi} = 0$. Subsequently performing the vector multiplication and integration implied in equation 8.26 we arrive at

$$\mathbf{H}_1 = -\mathbf{i}(I_{VV} - I_O)\Omega_p \sin \phi \cos \phi \sin \psi + \mathbf{j}(I_{VV} \sin^2 \phi + I_O \cos^2 \phi)\Omega_p$$
$$- \mathbf{k}(I_{VV} - I_O)\Omega_p \sin \phi \cos \phi \cos \psi \tag{8.29}$$

where I_O is the polar mass moment of inertia of the system about OO' (passing through the centre of the disc),

$$I_O = \frac{\rho b \pi R^4}{2}$$

and I_{VV} is the mass moment of inertia of the system about an axis passing through O, perpendicular to OO', i.e. parallel to the face of the disc as shown in Figure 8.5, which is in this case

$$I_{VV} = \frac{\rho b \pi R^4}{4} + \rho b \pi R^2 h^2$$

If we now consider equation 8.29 for the case where $\phi = 0$, we obtain

$$\mathbf{H}_1 = \mathbf{j} I_O \Omega_p$$

Likewise when $\phi = \pi/2$

$$\mathbf{H}_1 = \mathbf{j} I_{VV} \Omega_p$$

The reader should immediately see the compliance of the results of these two cases with the result of analysis of the spinning disc in Section 8.3.3.

Let us consider now the component of moment of momentum associated with the central spin of the body at Ω_s, namely $\mathbf{H}_2$. Consequently substituting $\phi' = \phi$ and $\psi' = \psi$ in equation 8.22 gives

$$\mathbf{H}_2 = I_O \Omega_s (\mathbf{i} \sin \phi \sin \psi + \mathbf{j} \cos \phi + \mathbf{k} \sin \phi \cos \psi) \tag{8.30}$$

Therefore the total moment of momentum, $\mathbf{H}$, of the spinning disc undergoing uniform precession at constant notation at the specified orientation is

$$\mathbf{H} = \mathbf{H}_1 + \mathbf{H}_2$$

i.e.

$$\begin{aligned}
\mathbf{H} = &\ \mathbf{i}[I_O \Omega_s \sin \phi \sin \psi - (I_{VV} - I_O)\Omega_p \sin \phi \cos \phi \sin \psi] \\
&+ \mathbf{j}[I_O \Omega_s \cos \phi + (I_{VV} \sin^2 \phi + I_O \cos^2 \phi)\Omega_p] \\
&+ \mathbf{k}[I_O \Omega_s \sin \phi \cos \psi - (I_{VV} - I_O)\Omega_p \sin \phi \cos \phi \cos \psi]
\end{aligned} \tag{8.31}$$

(b) $\phi' = \pi/2$, $\psi' = \psi - \pi/2$. In this case the line OO' and the local axis O'Z' are perpendicular, i.e. OO' is parallel to the face of the disc as shown in Figure 8.6, remembering that the axis O'X' is maintained parallel to the plane enclosed by the fixed reference axes OX and OZ.

Once again, as in (a), let us first derive the component of $\mathbf{H}$ due to the uniform precession only, i.e. $\mathbf{H}_1$. Then from equations 8.17, 8.23 and 8.24 we have

$$\begin{aligned}
\boldsymbol{\delta}_{OS} = &\ \mathbf{i}(h \sin \phi \sin \psi + r \cos \theta \sin \psi) + \mathbf{j}(h \cos \phi + r \sin \theta) \\
&+ \mathbf{k}(h \sin \phi \cos \psi + r \cos \theta \cos \psi)
\end{aligned} \tag{8.32}$$

and

$$\mathbf{v}_{OS} = \mathbf{i}\Omega_p (h \sin \phi + r \cos \theta) \cos \psi - \mathbf{k}\Omega_p (h \sin \phi + r \cos \theta) \sin \psi \tag{8.33}$$

Substituting equations 8.32 and 8.33 in equation 8.26 gives

$$\begin{aligned}
\mathbf{H}_1 = &\ -\mathbf{i}\Omega_p (Mh^2 \sin \phi \cos \phi \sin \psi) + \mathbf{j}\Omega_p (Mh^2 \sin^2 \phi + I_{DD}) \\
&- \mathbf{k}\Omega_p (Mh^2 \sin \phi \cos \phi \cos \psi)
\end{aligned} \tag{8.34}$$

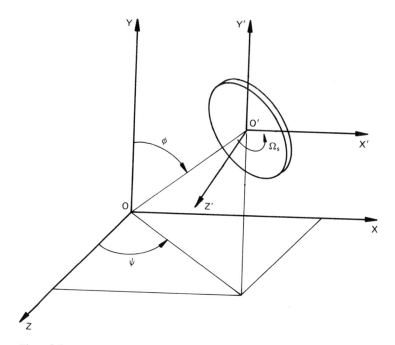

Figure 8.6

where $M = \rho b \pi R^2$, the mass of the disc, and $I_{DD} = \rho b \pi R^4/4$, the mass moment of inertia of the disc about its diagonal axis through O'.

Consider the case where $\phi = 0$, for which case we have

$$\mathbf{H}_1 = \mathbf{j}\Omega_p I_{DD}$$

Similarly, when $\phi = \pi/2$, we have

$$\mathbf{H}_1 = \mathbf{j}\Omega_p(Mh^2 + I_{DD}) = \mathbf{j}\Omega_p I_{YY}$$

where I_{YY} is the mass moment of inertia of the system about the axis OY.

Now for the component of **H** due only to the disc central axis spin at Ω_s, i.e. $\mathbf{H}_2$, we substitute $\phi' = \pi/2$ and $\psi' = \psi - \pi/2$ in equation 8.22, i.e.

$$\mathbf{H}_2 = I_0\Omega_s(-\mathbf{i}\cos\psi + \mathbf{k}\sin\psi) \tag{8.35}$$

and, as before, $\mathbf{H} = \mathbf{H}_1 + \mathbf{H}_2$.

(c) $\phi' = \phi - \pi/2$, $\psi' = \psi$. In this case, the orientation of the disc is as shown in Figure 8.7 where the *whole* face of the disc is maintained at an angle of $\pi/2 - \phi$ with the plane bounded by the reference axes OX and OZ.

Once again we start by deriving the component $\mathbf{H}_1$. From equations 8.17, 8.23 and 8.24 we have

$$\delta_{os} = \mathbf{i}(h\sin\phi\sin\psi + r\cos\theta\cos\psi - r\sin\theta\sin\phi\sin\psi) + \mathbf{j}(h\cos\phi - r\sin\theta\cos\phi)$$
$$+ \mathbf{k}(h\sin\phi\cos\psi - r\cos\theta\sin\psi - r\sin\theta\sin\phi\cos\psi) \tag{8.36}$$

and

$$\mathbf{v}_{os} = \mathbf{i}\Omega_p(h\sin\phi\cos\psi - r\cos\theta\sin\psi - r\sin\theta\sin\phi\cos\psi)$$
$$+ \mathbf{k}\Omega_p(-h\sin\phi\sin\psi - r\cos\theta\cos\psi + r\sin\theta\sin\phi\sin\psi) \tag{8.37}$$

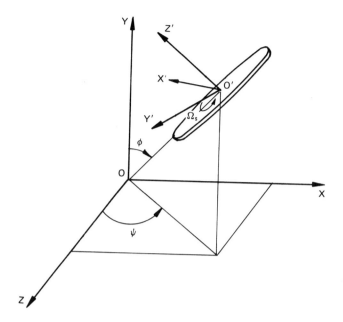

Figure 8.7

and substituting in equation 8.26 gives

$$\mathbf{H}_1 = -\mathbf{i}\Omega_p[Mh^2 + I_{DD}]\cos\phi\sin\phi\sin\psi + \mathbf{j}\Omega_p[I_{DD} + (Mh^2 + I_{DD})\sin^2\phi] \\ - \mathbf{k}\Omega_p[Mh^2 + I_{DD}]\cos\phi\sin\phi\cos\psi \quad\quad (8.38)$$

Now, when $\phi = 0$,

$$\mathbf{H}_1 = \mathbf{j}\Omega_p I_{DD}$$

and when $\phi = \pi/2$

$$\mathbf{H}_1 = \mathbf{j}\Omega_p(Mh^2 + 2I_{DD}) = \mathbf{j}\Omega_p(Mh^2 + I_0) \quad\quad \text{for a disc}$$

Now by substituting $\phi' = \phi - \pi/2$ and $\psi' = \psi$ in equation 8.22 we have

$$\mathbf{H}_2 = I_0\Omega_s(-\mathbf{i}\cos\phi\sin\psi + \mathbf{j}\sin\phi - \mathbf{k}\cos\phi\cos\psi) \quad\quad (8.39)$$

and once again

$$\mathbf{H} = \mathbf{H}_1 + \mathbf{H}_2$$

8.4   Gyroscopes and gyroscopic torque

A gyroscope can be defined basically as a solid body capable of rotating at a high angular velocity about an axis passing through a fixed point. More specifically, however, the usual form of a gyroscope is a flywheel so mounted that whilst spinning at a high speed of rotation its axis of rotation can turn in any direction within a fixed coordinate system usually by means of double gimbals called a *Cardan suspension* (see Figure 8.8).

This system is similar to that described in Figure 8.5 of Section 8.3.4, with $h = 0$.

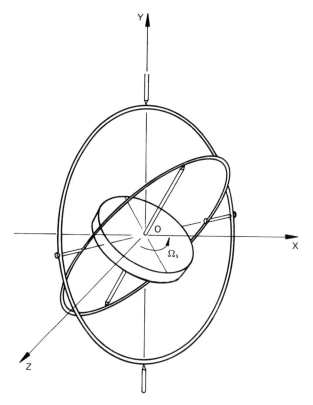

Figure 8.8

Therefore, setting $h = 0$ in equation 8.31, noting that $I_{YY} = I_0/2$, the moment of momentum, **H**, reduces to

$$\mathbf{H} = \mathbf{i}I_0[\Omega_s + \tfrac{1}{2}\Omega_p\cos\phi]\sin\phi\sin\psi + \mathbf{j}I_0[\Omega_s\cos\phi + \Omega_p(\tfrac{1}{2}\sin^2\phi + \cos^2\phi)]$$
$$+ \mathbf{k}I_0[\Omega_s + \tfrac{1}{2}\Omega_p\cos\phi]\sin\phi\cos\psi \tag{8.40}$$

and the rate of change of moment of momentum is

$$d\mathbf{H}/dt = I_0(\Omega_s\Omega_p + \tfrac{1}{2}\Omega_p^2\cos\phi)(\sin\phi)(\mathbf{i}\cos\psi - \mathbf{k}\sin\psi) \tag{8.41}$$

Also, in compliance with the relationship described by equation 8.16, the right-hand side of equation 8.41 represents the instantaneous net moment (or torque) necessary to be applied at O in order to achieve this steady change in ψ and constant value of ϕ. Furthermore, in practical mechanisms of this type, friction is so low as to be considered negligible; the net torque described by equation 8.41 is therefore, for all practical purposes, the torque required to sustain the motion. Conversely, when motion of the axis of spinning is induced, the gyroscope will give rise to an inertia torque, equal in magnitude *but opposed in sense*, to the torque described by equation 8.41.

In the present context, this inertia torque is given a special name, i.e. the gyroscopic torque, $\mathbf{T_g}$, and can be calculated from equation 8.41 with the signs reversed, i.e.

$$\mathbf{T_g} = -\,d\mathbf{H}/dt$$
$$= I_0(\Omega_s\Omega_p + \tfrac{1}{2}\Omega_p^2\cos\phi)(\sin\phi)(-\mathbf{i}\cos\psi + \mathbf{k}\sin\psi) \tag{8.42}$$

At this stage, let us consider some specific cases of this gyroscopic torque described above, remembering that we are considering only the case of uniform precession and constant nutation.

8.4.1   Uniform precession at an angle of nutation $\phi = 0°$

In this case, the polar axis of the flywheel is parallel to the OY axis and it is in effect spinning at a speed of $(\Omega_s + \Omega_p)$ about OY as shown in Figure 8.9. Substituting $\phi = 0$ in equation 8.42 gives, for this particular case,

$$\mathbf{T_g} = 0$$

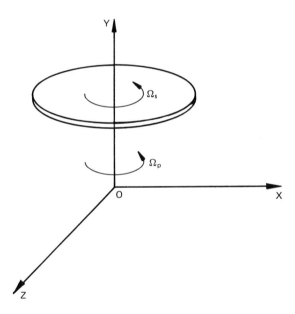

Figure 8.9

8.4.2   Uniform precession at an angle of nutation $\phi = 90°$

In this case, the spinning flywheel is rotating at a constant precessional angular velocity of Ω_p around the OY axis as shown in Figure 8.10. Furthermore, since $\phi = 90°$, then the axis of spinning O'Z' is always parallel to and on the plane bound by the axes OX and OZ. Hence, substituting $\phi = 90°$ in equation 8.42, the gyroscopic torque $\mathbf{T_g}$ is

$$\mathbf{T_g} = I_0\Omega_s\Omega_p(-\mathbf{i}\cos\psi + \mathbf{k}\sin\psi)$$

Hence the magnitude of $\mathbf{T}_g$ is

$$|\mathbf{T}_g| = I_0 \Omega_s \Omega_p$$

and, as a vector, acts along the local O'X' axis in a direction towards O' (see Figure 8.10). Therefore, translating the gyroscopic torque from a vector into its physical significance, by means of the right-hand corkscrew rule, we have a situation as illustrated in Figure 8.10.

The particular form of gyroscope we have investigated so far has, in modified forms, numerous specific applications such as the gyrostabilizer and the gyrocompass, the study of which are beyond the scope of this text. However, more generally, a mechanical system containing rotating members that are required to move through general space will be subjected to gyroscopic torques, e.g. aircraft, motor cars and ships, and perhaps the most obvious example—the common bicycle. Therefore in the course of the design of such systems, the inevitable presence of the gyroscopic torque must not only be appreciated, but also quantitatively accounted for in the design calculations.

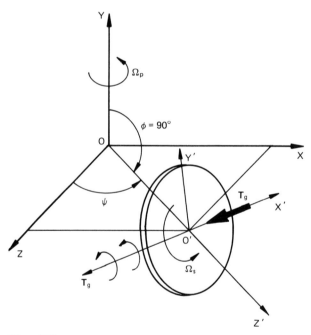

Figure 8.10

Example 8.1

A transverse engine automobile (crankshaft parallel with axles) is required to travel around an unbanked curve of 120 m mean radius. Each of the four road wheels has a polar mass moment of inertia of 4 kg m^2 and an effective diameter of 500 mm. The rotating parts of the engine have a polar mass moment of inertia of $2.5\,\text{kg}\,\text{m}^2$ and rotate in the same sense as the road wheels. The gear ratio, engine to road wheels is 1.5:1. The vehicle has a mass of 800 kg and its centre of gravity is 500 mm above the road. The width of the track of the vehicle is 1.5 m. Estimate the limiting speed of the vehicle around the curve for all wheels to maintain contact with the road surface. Assume that the vehicle does not slide, and that the difference in rotational speed between the outside and nearside wheels is negligible.

SOLUTION

Figure 8.11 shows a free body diagram of the vehicle as viewed from the rear. For convenience the vehicle is shown to be just passing the OX axis of the fixed reference axes system, i.e. $\psi = 90°$. Also we deduce that the direction of rotation of the road wheels (and engine) are opposite to that assumed throughout the foregoing analysis, i.e. $\Omega_s = -\Omega_s$ in this case. Also, if the constant precession angular velocity is Ω_p, then for the road wheels

$$\Omega_s = \frac{R\Omega_p}{r} = 480\Omega_p$$

where r, the radius of the road wheels, is 250 mm; and for the engine

$$\Omega_s = 1.5 \times 480\Omega_p = 720\Omega_p$$

Therefore, by differentiating equation 8.31 with respect to time and reversing the

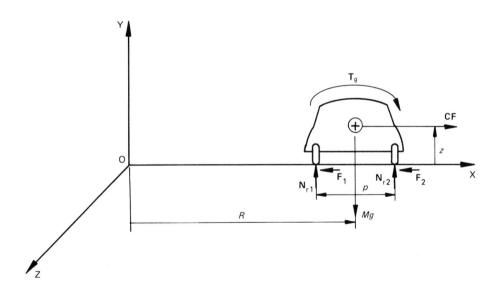

Figure 8.11

signs, for both ψ and ϕ set at $90°$, the gyroscopic torque, $\mathbf{T}_g$, is

$$\begin{aligned}
\mathbf{T}_g &= -\mathbf{k}[(I_0\Omega_s\Omega_p)_{\text{wheels}} + (I_0\Omega_s\Omega_p)_{\text{eng}}]\\
&= -\mathbf{k}[(4 \times 4 \times 480\Omega_p^2) + (2.5 \times 720\Omega_p^2)]\\
&= -\mathbf{k}\,9480\,\Omega_p^2
\end{aligned}$$

and acts in the sense shown in Figure 8.11. Now for all wheels to *just* remain in contact with the ground, $N_{r_1} = F_1 = 0$; and for no sliding to occur, $|\mathbf{CF}| = |\mathbf{F}_2| \leqslant \mu_s|\mathbf{N}_{r_2}|$, which we will assume to be the case. Therefore, for the vehicle to maintain stability,

$$Mgp/2 = |\mathbf{T}_g| + z|\mathbf{CF}|$$

i.e.

$$800 \times 9.81 \times 0.75 = 9480\Omega_p^2 + 800\Omega_p^2 \times 120 \times 0.5$$

giving

$$\Omega_p = 0.32\,\text{rad/s}$$

Therefore, maximum forward speed of vehicle,

$$R \times \Omega_p = 38.4\,\text{m/s}$$

The reader may wish to verify that, if the gyroscopic torque was not taken into account in the above calculations, the maximum forward speed of the vehicle would have been calculated as $42\,\text{m/s}$, i.e. approximately 9% higher than that previously calculated.

Problems

8.1 A rotor of polar mass moment of inertia I_0, rotating at Ω_s rad/s, is situated in a moving spacecraft as shown in Figure 8.12. If the spacecraft has rotational

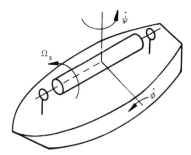

Figure 8.12

speeds, with respect to a fixed axes system, of $\dot{\phi}$ and $\dot{\psi}$ as shown, calculate the magnitude of the torque transmitted to the frame supporting the rotor.

Answer: $I_0\Omega_s(\dot{\phi} + \dot{\psi})^{1/2}$.

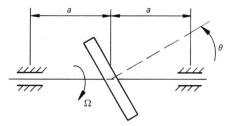

Figure 8.13

8.2   Figure 8.13 shows a solid disc inclined at an angle of θ to a rotating shaft. The mass and radius of the disc are M and R respectively. Derive an expression describing the magnitude of the reaction force at each of the bearings due to this inclination when the shaft rotates at Ω. Assume that the shaft is simply supported at the bearings.

Answer: $(MR^2/16a)\Omega^2 \sin 2\theta$.

8.3   An electric motor on board a ship is arranged with its rotor athwart the ship. Find the maximum load on its bearings due to gyroscopic action if the ship rolls with simple harmonic motion 30° on each side of the vertical and the time for one complete roll is 3.4 s. The mass of the rotor is 220 kg, its radius of gyration is 215 mm, the bearings are 1.1 m apart and the speed of the rotor is 3000 rev/min clockwise, viewed from the starboard side.

Answer: 2.81 kN.

8.4   A solo motor cycle, complete with rider, has a mass of 225 kg, the centre of gravity being 0.6 m above the ground level. The moment of inertia of each road wheel is 1 kg m² and the rolling diameter is 0.6 m. The engine crankshaft rotates, in the same sense as the wheels, at five times the speed of the wheels. The rotating parts of the engine have a moment of inertia of 0.2 kg m². Determine the heel-over angle required when the unit is travelling at 100 km/h around a curve of radius 60 m.

Answer: 54° 36′.

8.5   Referring to Figure 8.14, the axis OZ of a thin-walled right circular cone of base radius r makes a constant angle ϕ with the vertical OY′. It rotates about OY′ with a constant precession angular velocity $\dot{\psi} = \Omega_p$. At the same time the cone spins about its own axis OZ with a constant angular velocity Ω_s.

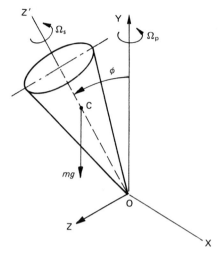

Figure 8.14

Given that

$OC = 2r$

$I_{ox} = I_{ox} = 19mr^2/4$

$I_{oz'} = mr^2/2$

show that

$$\Omega_p^2 + \left(\frac{2\Omega_s \sec\phi}{17}\right)\Omega_p + \frac{8g \sec\phi}{17r} = 0$$

9
Vibration of a single degree of freedom system

9.1 Introduction

In mechanical engineering there are many examples of cyclic or periodic motion, i.e. motion which is repeated in equal intervals of time.

Perhaps the simplest example is the pendulum, which when displaced from its equilibrium position and subsequently released will perform oscillatory motion of a given frequency and time interval; the restoring effect in such a device is, of course, gravity. If, however, a mechanical system which possesses both mass and elasticity is deflected from its equilibrium position and then released, the ensuing oscillatory motion is referred to as **vibration**. The frequency of the resulting motion will correspond to one of the so-called **natural frequencies** of the system; these are the frequencies the system will adopt when influenced only by the local parameters of mass and stiffness, i.e. in the absence of external effects. Associated with each of these natural frequencies is a normal mode shape which depicts the manner of movement of the system as it performs periodic motion. The amplitude of the oscillatory motion will, for all practical systems, diminish with time due to the dissipation of kinetic energy resulting from molecular friction within the material providing the elasticity, or alternatively by some form of externally applied damping.

In later chapters the more complex vibratory motion associated with mechanical systems which consists of distributed mass and elasticity, as well as those having many concentrated masses, will be examined in detail.

In this chapter, however, we shall confine our analysis to the simplest configuration, namely, the single degree of freedom system which requires only a single coordinate to define the displacement of the system at any instant; this single coordinate may be the line displacement in a rectilinear system or the rotational displacement in an angular system. In many practical cases the oscillatory motion will be a combination of both rectilinear and angular components and it will be helpful in such instances for the reader to consult once again the concept of a dynamically equivalent system outlined in Chapter 1. By this means, apparently complicated rectilinear/angular systems may, conveniently, be resolved as much simpler systems, where the resulting motion is purely rectilinear or angular. In the vibration analysis of mechanical systems, the systems generally comprise mass, stiffness and damping elements. The characteristics of damping will be investigated in Section 9.4. With respect to stiffness, however, very often systems contain a number of stiffness elements arranged either in series or in parallel, and we shall begin by considering how such arrangements can be reduced to a single stiffness element.

9.2   Series and parallel stiffness arrangements

9.2.1   Stiffnesses in series

Consider the three linearly elastic springs of respective stiffnesses K_1, K_2 and K_3, which are arranged in the manner shown in Figure 9.1a; the springs are said to be connected in series. If such a combination were present in a vibrating mass/elastic system, it might be more convenient to replace this with a single equivalent spring having a stiffness K_e (Figure 9.1b), such that under the action of the same applied force, F, the displacement δ of both systems will be the same. For the actual system, let the deflections of K_1, K_2 and K_3, be δ_1, δ_2 and δ_3 respectively, such that

$$\delta = \delta_1 + \delta_2 + \delta_3 \tag{9.1}$$

Now since the force F is constant through each stiffness

$$\delta_1 = \frac{F}{K_1} \qquad \delta_2 = \frac{F}{K_2} \qquad \delta_3 = \frac{F}{K_3} \tag{9.2}$$

and for the equivalent system

$$\delta = F/K_e \tag{9.3}$$

Substituting equations 9.2 and 9.3 into equation 9.1 gives

$$\frac{1}{K_e} = \frac{1}{K_1} + \frac{1}{K_2} + \frac{1}{K_3}$$

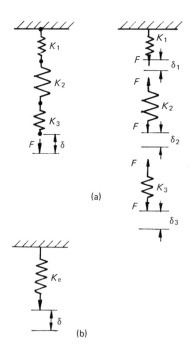

(a)

(b)

Figure 9.1

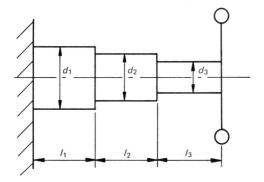

Figure 9.2

Similarly, if we had n stiffnesses in series, the equivalent stiffness would be calculated from

$$\frac{1}{K_e} = \frac{1}{K_1} + \frac{1}{K_2} + \frac{1}{K_3} + \cdots + \frac{1}{K_n} \tag{9.4}$$

The torsional analogy of linear stiffnesses in series is a stepped diameter shaft such as that shown in Figure 9.2. Now from basic theory for the twisting of a uniform cross-section bar of length l we have the relationship

$$\text{Torsional stiffness of shaft} = K = \frac{GJ_0}{l} \tag{9.5}$$

where $J_0 = \pi d^4/32$, the polar second moment of area of the shaft section (circular), and G is the modulus of rigidity of the shaft material.

In cases such as this it is often convenient to replace the actual system by an equivalent system comprising a shaft of the same material (same value of G) of length l_e and diameter d_e. Therefore, substituting equation 9.5 into the general equation 9.4 gives

$$\frac{l_e}{d_e^4} = \frac{l_1}{d_1^4} + \frac{l_2}{d_2^4} + \frac{l_3}{d_3^4} \tag{9.6}$$

Very often, for convenience, the value of d_e is assigned the value of either d_1, d_2 or d_3.

9.2.2 Stiffnesses in parallel

Consider, now, the alternative case where the three linearly stiffnesses are connected in parallel to a rigid frame, such that a common deflection, δ, is produced when a force F is applied to the frame as shown in Figure 9.3. If the individual forces acting through each stiffness are f_1, f_2 and f_3, as shown, then

$$F = f_1 + f_2 + f_3 \tag{9.7}$$

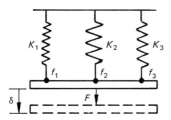

Figure 9.3

and, since the deflection of each stiffness is the same,

$$f_1 = K_1\delta \qquad f_2 = K_2\delta \qquad f_3 = K_3\delta \tag{9.8}$$

Substituting equations 9.8 into equation 9.7 gives

$$F = (K_1 + K_2 + K_3)\delta \tag{9.9}$$

Equation 9.9 indicates that the three parallel stiffnesses can be replaced by an equivalent stiffness of value equal to the sum of the three stiffnesses. Therefore, for a system containing n parallel stiffnesses, the equivalent stiffness value, K_e, is

$$K_e = K_1 + K_2 + K_3 + \cdots + K_n \tag{9.10}$$

Figure 9.4 shows an example of parallel torsional stiffnesses where a system of concentric shafts controls the angular position of the rotor. For any angular movement of the rotor, the angle of twist in each shaft is the same, therefore the stiffness of the system (torque at rotor/angular deflection) will be the sum of the torsional stiffnesses of the three shafts.

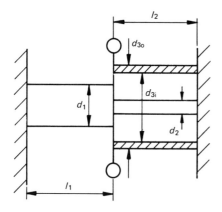

Figure 9.4

9.3 Free vibration of an undamped system

Consider the single mass/stiffness arrangement shown in Figure 9.5. Let y_0 denote the magnitude of the downwards displacement of the end of the spring attached to

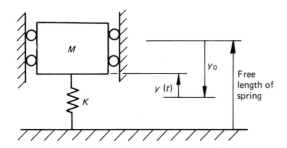

Figure 9.5

the mass due to the gravitational force of the mass, and $y(t)$ as the subsequent instantaneous upward displacement of the mass from this static equilibrium position as shown.

If we apply Newton's Second Law of Motion to the mass, denoting all forces and displacements upwards to be positive and vice versa, we obtain

$$-K[y(t) - y_0] - Mg = M \, d^2 y(t)/dt^2 = M \, \ddot{y}(t)$$

and since $Mg = Ky_0$, the above equation simplifies to

$$M \, \ddot{y}(t) + K \, y(t) = 0 \tag{9.11}$$

The solution to the above linear second-order differential equation with constant coefficients may be obtained in many ways. The most basic method of solution commences with assuming that $y(t) = A\exp(\alpha t)$, where A and α are constants and t is general time. Therefore substituting for $y(t)$ in equation 9.11 gives

$$M\alpha^2 + K = 0$$

giving

$$\alpha = \pm \sqrt{(-K/M)} = 0 \pm i\sqrt{(K/M)}$$

where i is the complex conjugate, $\sqrt{-1}$. On the basis of the above root values of α, the general solution to equation 9.11 is

$$y(t) = A_1 \cos \omega t + A_2 \sin \omega t \tag{9.12}$$

where A_1 and A_2 are arbitrary constants of integration, normally determined from the specified values of $y(t)$ and $\dot{y}(t)$ at $t = 0$ (initial conditions) and $\omega = \sqrt{(K/M)}$ is the **undamped natural frequency** of the system. Alternatively, equation 9.12 can be expressed in the more convenient form

$$y(t) = A \cos(\omega t + \varepsilon) \tag{9.13}$$

where A and ε are constants of integration, once more calculated from the initial conditions.

As has been stated in Section 9.1, the natural frequency is the frequency the system will describe if disturbed from the equilibrium position and subsequently released, with no external forces influencing the motion. The number of natural frequencies which a mass/elastic system has is equal to the number of degrees of freedom the system possesses, i.e. the number of independent coordinates required to fully specify the motion of the system. Since the particular system being examined has a single degree of freedom, it will, consequently, have a single natural frequency, namely, ω.

The values of A and ε will, of course, depend entirely upon the manner in which vibration is initiated; for example, if the mass is slowly displaced from the equilibrium position and subsequently released, the initial conditions will be such that $\dot{y}(0) = 0$. Thus differentiating equation 9.13 with respect to time and equating to zero for $t = 0$ gives

$$0 = -\omega A \sin(\varepsilon)$$

From the above equation there are three possible solutions, namely A is zero, ε is zero, or A and ε are both zero. If A is zero, equation 9.13 would suggest zero vibration for all time. Therefore the only non-trivial solution is that $\varepsilon = 0$ (or $n\pi$ where n is an integer). Therefore, for this particular example,

$$y(t) = A \cos \omega t$$

where A represents the amplitude of vibratory motion of the body. The variation of $y(t)$ in terms of t is shown in Figure 9.6 where the vertical axis represents displacement and the horizontal axis corresponds to time. It will be observed that A, the amplitude of vibration, continued undiminished, since there is no energy dissipation mechanism within the system, i.e. the system is undamped. In Section 9.4 the effects of damping upon the vibratory motion of a single degree of freedom system will be studied.

Although the foregoing analysis was related to a single mass/spring arrangement, vibrating in a straight line, the equations derived, i.e. equations 9.12 and 9.13, also apply to single degree of freedom torsional systems, in which case the mass M is replaced by I, the polar mass moment of inertia, and the linear stiffness K (in units of N/m) is replaced by an equivalent angular stiffness (in units of N m/rad). For such a system the response would be expressed in terms of $\theta(t)$, the angular displacement of the system about the relevant axis of oscillation, as opposed to the linear displacement $y(t)$.

In some systems, however, the vibratory motion induced after initial disturbance may consist of both rectilinear and angular components; in such cases the complexity of the analysis may be greatly reduced by employing the method of dynamically equivalent systems, as detailed in Chapter 1. In this way the initially complex

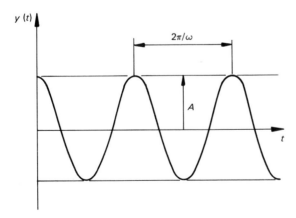

Figure 9.6

arrangement may be replaced by one where the vibratory motion is either entirely rectilinear or alternatively angular. The following example illustrates clearly the technique of dynamical equivalence.

Example 9.1

The elements of a vibration recorder are shown in Figure 9.7. The rod OB, which is pivoted at O, has a mass of 1.2 kg and a radius of gyration of 58 mm about O. The controlling spring attached to the rod at C has a stiffness of 1400 N/m and the block at B has a mass of 1.8 kg but its radius of gyration about its own centre of mass can be neglected.

(a) Determine the undamped natural frequency of free vibration of the system.
(b) It is required to raise the natural frequency by 10% by changing either the value of the mass at B or the stiffness of the system. To satisfy this requirement determine
   (i) the change in value of the mass of the block at B,
   (ii) the stiffness of an additional spring acting on the lever and attached at G.

SOLUTION

(a) Since this problem represents a system describing combined rectilinear and angular motion, solution may be simplified by re-expressing it as a simple dynamically

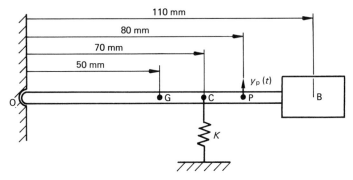

Figure 9.7

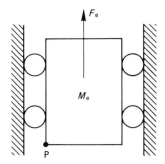

Figure 9.8

equivalent system where the motion is either totally rectilinear or totally angular. Let us re-express it as a dynamically equivalent mass M_e undergoing rectilinear motion at the point P under the action of a dynamically equivalent force F_e as shown in Figure 9.8. The reader will note that the system shown in Figure 9.7 is identical to the system analysed in Section 1.2.2 of Chapter 1 in the absence of the applied force F. Therefore, from the analysis contained in Section 1.2.2, we have

M_e = dynamically equivalent mass at P

$$= \frac{1.2 \times 58^2}{80^2} + 1.8 \times \left(\frac{110}{80}\right)^2$$

$$= 0.630 + 3.404 = 4.034\,\text{kg}$$

F_e = dynamically equivalent force at the point P

$$= -1400 \times \left(\frac{70}{80}\right)^2 y_p(t)$$

$$= -1071.875\,y_p(t)\,\text{N}$$

where $y_p(t)$ is the instantaneous vertical displacement of the point P from the static equilibrium position. Therefore we can write

$$-1071.875\,y_p(t) = 4.034\,\ddot{y}_p(t)$$

or

$$4.034\,\ddot{y}_p(t) + 1071.875\,y_p(t) = 0$$

If we compare the above equation with equation 9.11, we see that it represents the equation of motion pertaining to a simple mass/spring system similar to that of Figure 9.5 with $M = M_e$ = dynamically equivalent mass = 4.034 kg and $K = K_e$ = dynamically equivalent stiffness = 1071.875 N/m. Therefore, in this case,

$$\omega = \sqrt{\left(\frac{K_e}{M_e}\right)} = \sqrt{\left(\frac{1071.875}{4.034}\right)} = 16.3 \ \text{rad/s} = 2.59 \ \text{Hz}$$

(b) Required undamped natural frequency $= 1.1 \times 16.3 = 17.93\,\text{rad/s}$, i.e.

$$\omega^2 = 321.49 \quad \text{rad}^2/\text{s}^2$$

(i) Let the mass at B = M_b. Therefore the dynamically equivalent mass at P, M_e, becomes

$$M_e = 0.630 + 1.89 M_b \tag{1}$$

Now from the relationship $\omega^2 = 321.49 = 1071.875/M_e$, we find that $M_e = 3.334\,\text{kg}$. Therefore, from equation 1 we solve for M_b to give 1.43 kg.

(ii) In this case M_e remains at 4.034 kg but by adding an extra stiffness at G we in effect change the dynamically equivalent force at P to

$$F_e = -1071.875\,y_p(t) - K_g(50/80)^2\,y_p(t)$$

where K_g is the added stiffness at G, not to be confused with the unit mass, kg. Therefore the dynamically equivalent stiffness, K_e, at P is

$$K_e = (1071.875 + 0.39 K_g) \tag{2}$$

Also, from the relationship that $\omega^2 = 321.49 = K_e/M_e$, we find that $K_e = 1296.7\,\text{N/m}$. Therefore, from equation 2 we solve for K_g to obtain 576 N/m.

9.4   Free vibration of damped systems

In free undamped vibration, the mass/elastic system, having been set into vibratory motion, will continue without loss of energy, i.e. the vibration is considered to be an interchange, without loss, between two energy stores, namely, kinetic and potential (strain); however, all practical systems are influenced, to some degree, by external forces, and the vibratory motion will diminish with time as the total energy is dissipated in the form of heat or in some cases as radiated sound; such systems are referred to as **damped** systems, and the mechanism of damping may be fluid or solid molecular.

Fluid damping may be viscous or turbulent, although for ease of solution the former is generally assumed and the damping force taken as being proportional to velocity. Energy dissipation which is induced by the rubbing of dry surfaces, rather than lubricated ones, is referred to as Coulomb damping and the damping force, which is due to the kinetic friction between the contacting surfaces, is assumed to be constant.

Solid molecular or hysteretic damping results from friction between the molecules of the elastic material which is deformed during vibration.

For the purposes of this book, dynamic analysis of the vibratory response of damped mass/elastic systems will be confined to systems where damping is of the viscous type.

9.4.1   Vibratory response

Consider the arrangement shown in Figure 9.9 where the body of mass M is not only acted upon by the elastic force, $\dot{y}(t)$, but is additionally resisted by the force produced by the oil dashpot of magnitude $C\,y(t)$, acting in a direction opposite to the direction of the velocity vector of the body; C is the coefficient of viscous damping of the dashpot, which in itself is simply a diagrammatic representation of all resistance forces acting upon the body, which are directly proportional to the velocity of the body. Once again positive displacement and velocity are assumed to be in the direction indicated.

If we apply Newton's Second Law of Motion to the body, at a point where the displacement of the body from the static equilibrium position is $y(t)$, then, for a body

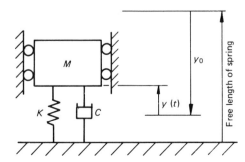

Figure 9.9

of constant mass, the governing equation of motion will be

$$-K\,y(t) - C\,\dot{y}(t) = M\,\ddot{y}(t)$$

or

$$M\,\ddot{y}(t) + C\,\dot{y}(t) + K\,y(t) = 0 \tag{9.14}$$

Again we have a linear, homogeneous second-order differential equation with constant coefficients. Before solving, it will be found more convenient to express equation 9.14 in the modified form

$$\ddot{y}(t) + 2\Delta\,\dot{y}(t) + \omega^2 y(t) = 0 \tag{9.15}$$

In this form $\Delta = C/2M$, the modified coefficient of damping, and $\omega = \sqrt{(K/M)}$, the undamped natural frequency of the system. Once again we commence by assuming $y(t)$ to be of the general form $y(t) = A\exp(\alpha t)$ and substituting in equation 9.15 gives

$$\alpha^2 + 2\Delta\alpha + \omega^2 = 0$$

giving

$$\alpha = -\Delta \pm \sqrt{(\Delta^2 - \omega^2)} \tag{9.16}$$

From the above relationship it is obvious that the solution for α and consequently the solution for $y(t)$ will depend on the relative values of Δ and ω. Three such relationships exist, namely $\Delta < \omega$, $\Delta > \omega$ and $\Delta = \omega$. We shall consider each of these cases in turn.

Case 1: $\Delta < \omega$
In this case, where the modified coefficient of viscous damping is numerically less than the undamped natural frequency of the system, the system is said to be **underdamped**. In this case, therefore, root values of α are, from equation 9.16,

$$\alpha_1 = -\Delta + i\sqrt{(\omega^2 - \Delta^2)} \qquad \text{and} \qquad \alpha_2 = -\Delta - i\sqrt{(\omega^2 - \Delta^2)}$$

where, once again, i is the complex conjugate, $\sqrt{-1}$, For such a case, let us define a new term,

$$\omega_{\mathrm{d}} = \sqrt{(\omega^2 - \Delta^2)} \tag{9.17}$$

which is the natural frequency of damped vibratory motion of the system, i.e. the frequency at which the system will freely vibrate after being disturbed from the static equilibrium position. For these forms of root values, the general solution of $y(t)$ is

$$y(t) = \exp(-\Delta t)(A_1\cos\omega_{\mathrm{d}}t + A_2\sin\omega_{\mathrm{d}}t)$$

where A_1 and A_2 are constants of integration. As was the case with the free undamped system, the vibratory motion of the system will be more readily appreciated if the response $y(t)$ is written in terms of a single circular function rather than the sum of two such functions, i.e.

$$y(t) = A\exp(-\Delta t)\cos(\omega_{\mathrm{d}}t + \varepsilon) \tag{9.18}$$

Both the constants A and ε will depend upon the initial conditions governing a specific system. The corresponding graphical representation of equation 9.18 is illustrated in Figure 9.10, where the response is seen to be that of an exponentially decaying harmonic wave, having a circular frequency ω_{d} and related periodic time $T = 2\pi/\omega_{\mathrm{d}}$, which is contained within the envelope of the curve described by $A\exp(-\Delta t)$.

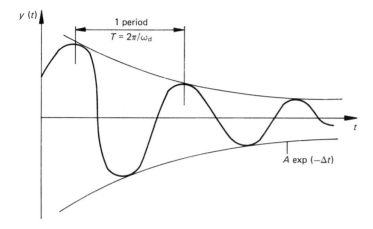

Figure 9.10

Case 2: $\Delta > \omega$

For this case the system is described as **overdamped**. Again, for mathematical convenience we introduce a further term, p, such that

$$p = \sqrt{(\Delta^2 - \omega^2)}$$

Substituting in equation 9.16 gives

$$\alpha_1 = -\Delta + p \qquad \text{and} \qquad \alpha_2 = -\Delta - p$$

in which case the solution of $y(t)$ is

$$y(t) = A_1 \exp[(-\Delta + p)t] + A_2 \exp[(-\Delta - p)t] \tag{9.19}$$

Since $\Delta > p$, $y(t)$ will always be an exponentially *decaying* function, reducing with time. The graphical representation of equation 9.19 is shown in Figure 9.11.

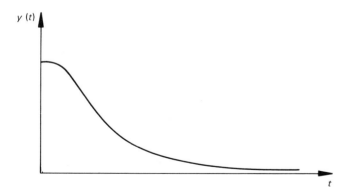

Figure 9.11

After being initially disturbed and subsequently released, the overdamped system returns to the equilibrium position in a slow logarithmic manner; at no time does the displacement pass through the rest position, i.e. no oscillatory motion takes place.

Case 3: $\Delta = \omega$

When the modified coefficient of damping is equal to the natural undamped frequency of vibration of the system, such a system is considered to be **critically damped**. From equation 9.16 we have, in this instance, the root values of α to be $-\Delta$, twice. In this case the general solution for $y(t)$ is

$$y(t) = \exp(-\Delta t)(A_1 t + A_2) \tag{9.20}$$

Once again, following an initial disturbance the critically damped system will return to the equilibrium position, exponentially. This degree of damping marks the minimum dampling required to ensure that no oscillation occurs.

The actual degree of damping present within a system is often expressed in terms of the critical damping condition, e.g. 1/4 critically damped means $\Delta = \Delta_c/4 = \omega/4$, where Δ_c is the modified critical damping coefficient. Alternatively, damping may be defined in non-dimensional form as

$$\Delta/\Delta_c = \xi = \text{non-dimensional damping ratio} \\ = \Delta/\omega \tag{9.21}$$

Generally, for critically damped systems $\xi = 1$, for overdamped systems $\xi > 1$, and for underdamped systems $\xi < 1$. Therefore, the vibratory response of an underdamped single degree of freedom system (equation 9.18) may be re-expressed in non-dimensional damping terms as

$$y(t) = A \exp(-\xi\omega t)\cos(\omega_d t + \varepsilon) \tag{9.22}$$

where

$$\omega_d = \omega\sqrt{(1 - \xi^2)} \tag{9.23}$$

9.4.2 Analysis of transient waveform

By far the most important of the three cases considered in terms of damped vibration, as applied to mechanical engineering, is the underdamped case, since only for this degree of damping will a transient oscillatory motion of the spring/mass system actually occur. With this in mind, let us take a closer look at the transient waveform that represents the response of the system when disturbed from its static equilibrium position. Such an analysis can facilitate the determination of the degree of damping present within a given system, no matter how complex that system may be. For the underdamped case, the solution was given as

$$y(t) = A \exp(-\Delta t)\cos(\omega_d t + \varepsilon) \tag{9.24}$$
$$= 0 \qquad \text{when } \cos(\omega_d t + \varepsilon) = 0$$

since $A = 0$ would produce no vibration at any time. Thus for zero vibratory amplitude we have

$$\omega_d t + \varepsilon = \pi/2 \pm n\pi \qquad \text{where } n \text{ is an integer}$$

and $t_{y=0}$, the times corresponding to this condition, would be given as

$$t_{y=0} = \frac{\pi/2 - \varepsilon \pm n\pi}{\omega_d} \tag{9.25}$$

which indicates that positions of zero vibratory amplitude repeat at regular intervals of π/ω_d s, i.e. every half period. To investigate the variation of linear velocity of the mass/elastic system, we differentiate equation 9.24 with respect to time and obtain

$$\dot{y}(t) = -A \exp(-\Delta t)[\Delta \cos(\omega_d t + \varepsilon) + \omega_d \sin(\omega_d t + \varepsilon)] \tag{9.26}$$

For maximum vibratory amplitude, $\dot{y}(t) = 0$, which implies that the terms within the square brackets must be zero. Consequently

$$\Delta \cos(\omega_d t + \varepsilon) = -\omega_d \sin(\omega_d t + \varepsilon)$$

or

$$\tan(\omega_d t + \varepsilon) = \frac{-\Delta}{\omega_d}$$

i.e.

$$\omega_d t + \varepsilon = \arctan\left(\frac{-\Delta}{\omega_d}\right) \pm n\pi$$

It follows, therefore, that positions of maximum vibratory amplitude, corresponding to zero vibratory velocity of the system, will be given by

$$t_{\dot{y}=0} = \frac{\arctan(-\Delta/\omega_d) - \varepsilon \pm n\pi}{\omega_d} \tag{9.27}$$

which again corresponds to a repetition every half period. These relationships are illustrated in Figure 9.12.

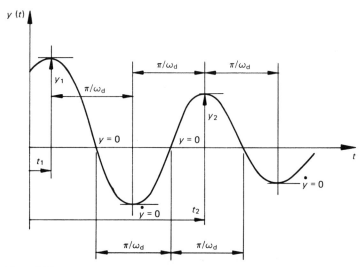

Figure 9.12

Note that although the time interval between consecutive peaks is a half period, as is the time interval between consecutive zeros, the time interval between an adjacent peak and zero is not a quarter period, although for very light damping the error incurred in making such an assumption will be small.

Successive maximum amplitudes have a logarithmic relationship which may be demonstrated as follows.

At time t_1, let the amplitude of vibration of the system be y_1 as shown in Figure 9.12; and at time t_2, let the amplitude be y_2, such that y_1 and y_2 are successive positive maxima, and the time interval between t_1 and t_2, represents exactly one period of vibration, i.e. $t_2 - t_1 = T = 2\pi/\omega_d$. Substituting in equation 9.24 gives the relationships

$$y_1 = A \exp(-\Delta t_1) \cos(\omega_d t_1 + \varepsilon)$$

and

$$y_2 = A \exp(-\Delta t_2) \cos(\omega_d t_2 + \varepsilon) \qquad (9.28)$$

Dividing these and recalling that $t_2 = t_1 + 2\pi/\omega_d$,

$$\frac{y_1}{y_2} = \exp[\Delta(t_2 - t_1)] = \exp[\Delta 2\pi/\omega_d] \qquad (9.29)$$

In general, equation 9.29 may be written in the form

$$\frac{y_i}{y_{i+q}} = \exp(q\Delta 2\pi/\omega_d) \qquad \text{where } q = 1, 2, 3 \cdots$$

or

$$\frac{y_i}{y_{i+q}} = -\exp(q\Delta 2\pi/\omega_d) \qquad \text{where } q = \tfrac{1}{2}, \tfrac{3}{2}, \tfrac{5}{2} \cdots$$

(9.30)

The logarithmic relationship between consecutive cycles may be expressed in terms of the *logarithmic decrement*, δ, which is defined as

$$\delta = \ln\left(\frac{y_1}{y_2}\right) = \frac{2\pi\Delta}{\omega_d} \qquad (9.31)$$

or in terms of the non-dimensional damping factor, ξ, the logarithmic decrement will be

$$\delta = \frac{2\pi\xi}{\sqrt{(1 - \xi^2)}} \qquad (9.32)$$

Therefore, in practice, the degree of damping present in a system can be estimated by disturbing the system from its static equilibrium position and recording the transient response on some suitable recorder. Consequently the logarithmic decrement can be measured, enabling the degree of viscous damping present within the system to be quantified.

9.5 Forced vibrations

In the previous sections of this chapter we considered the free response of a single mass/elastic system, i.e. in the absence of any external influences other than the elastic and damping forces. In many practical systems, however, in addition to these effects additional time-dependent forces may be applied, directly or indirectly, to the mass of the system and in such cases the vibratory response will change considerably. The time-dependent force or forces may be represented by a step or ramp input or be random in nature. They may also be periodic but not necessarily harmonic, although such forces may be reduced to a series of harmonic functions using, for example, Fourier harmonic analysis, and the response of the system thereby obtained for each salient harmonic component.

9.5.1   Steady-state vibratory response

In this section we shall examine the response of a single degree of freedom system which is acted upon by a single harmonic forcing component $p(t) = P \cos \Omega t$, where the frequency of application is Ω rad/s and the peak value of the force is P; the arrangement is illustrated in Figure 9.13. Viscous damping within the system is represented by the oil dashpot, which has a coefficient of viscous damping C.

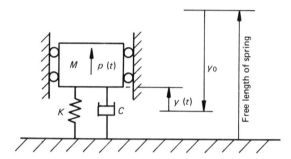

Figure 9.13

As before, we commence analysis by applying Newton's Second Law of Motion to the mass M, which gives

$$P \cos \Omega t - C \dot{y}(t) - K y(t) = M \ddot{y}(t) \tag{9.33}$$

or

$$M \ddot{y}(t) + C \dot{y}(t) + K \dot{y}(t) = P \cos \Omega t$$

and dividing throughout by M, remembering that $C/M = 2\xi\omega$, $K/M = \omega^2$, and $P/M = (K/M)(P/K) = \omega^2 Y_0$, where $Y_0 = P/K$, gives

$$\ddot{y}(t) + 2\xi\omega \dot{y}(t) + \omega^2 y(t) = \omega^2 Y_0 \cos \Omega t \tag{9.34}$$

Part of the solution to equation 9.34 relates to the free damped vibration of the system and is obtained by utilizing the homogeneous equation 9.15 previously examined. In addition there will be a particular solution which relates to the steady oscillation of the system at the impressed frequency of excitation, Ω. Thus the complete solution to equation 9.34 is the sum of transient and steady-state components and without a rigorous mathematical analysis the solution, assuming the underdamped case, may be shown to produce the relationship

$$y(t) = A \exp(-\xi\omega t) \cos(\omega_d t + \varepsilon) + Y \cos(\Omega t + \phi)$$
$$= \text{transient component} \quad + \text{steady-state component} \tag{9.35}$$

where Y represents the maximum amplitude of the steady-state component of the vibration and ϕ is the phase angle between the applied force and the ensuing steady-state vibratory displacement of the mass.

The negative exponent of the transient part of the general solution, being time dependent, will of course diminish as the vibration progresses. The response of the system in graphical terms is shown in Figure 9.14, for the case where the damped natural frequency of the system is much greater than the frequency of excitation.

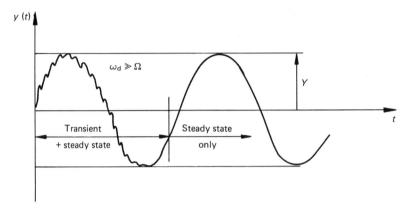

Figure 9.14

The transient part of equation 9.35 is of great importance where shock or impulsive forces act upon the system, but may be neglected in preference to the steady-state solution where long-term excitation of the system exists which may, if of significant magnitude, induce fatigue failure of individual components or indeed of complete systems.

Let us therefore concentrate our analysis on the steady-state component of equation 9.35. At this stage it will be more convenient, from the mathematical aspect, to represent the steady-state response of the system in complex notation, i.e.

$$y(t) = Y\underline{/\phi}\cos\Omega t = (Y_r + \mathrm{i}\,Y_i)\cos\Omega t \tag{9.36}$$

where Y_r is the real component of the steady-state response $(= Y\cos\phi)$, Y_i is the imaginary component of the steady-state response $(= Y\sin\phi)$, and i is the complex conjugate $\sqrt{-1}$.

Therefore, substituting for $y(t)$ from equation 9.36 into equation 9.34 gives

$$-\Omega^2(Y_r + \mathrm{i}\,Y_i)\cos\Omega t - 2\xi\omega\Omega(Y_r + \mathrm{i}\,Y_i)\sin\Omega t + \omega^2(Y_r + \mathrm{i}\,Y_i)\cos\Omega t = \omega^2 Y_0\cos\Omega t \tag{9.37}$$

Now,

$$
\begin{aligned}
-2\xi\omega\Omega(Y_r + \mathrm{i}\,Y_i)\sin\Omega t &= -2\xi\omega\Omega(Y_r + \mathrm{i}\,Y_i)\cos(\Omega t - \pi/2)\\
&= -2\xi\omega\Omega(-\mathrm{i})(Y_r + \mathrm{i}\,Y_i)\cos\Omega t\\
&= 2\xi\omega\Omega(-Y_i + \mathrm{i}\,Y_r)\cos\Omega t
\end{aligned}
$$

Therefore equation 9.37 becomes

$$-\Omega^2(Y_r + \mathrm{i}\,Y_i) + 2\xi\omega\Omega(-Y_i + \mathrm{i}\,Y_r) + \omega^2(Y_r + \mathrm{i}\,Y_i) = \omega^2 Y_0$$

and dividing throughout by ω^2 gives

$$-f^2(Y_r + i\,Y_i) + 2\xi f(-Y_i + i\,Y_r) + (Y_r + i\,Y_i) = Y_0 \tag{9.38}$$

where $f = \Omega/\omega$. If we now equate real to real and imaginary to imaginary terms we obtain the simultaneous equation

$$
\begin{aligned}
(1 - f^2)Y_r - 2\xi f Y_i &= Y_0 \\
2\xi f Y_r + (1 - f^2)Y_i &= 0
\end{aligned}
\tag{9.39}
$$

and solving for Y_r and Y_i gives

$$Y_r = \frac{(1 - f^2)Y_0}{D} \quad \text{and} \quad Y_i = \frac{-2\xi f Y_0}{D}$$

where $D = (1 - f^2)^2 + 4\xi^2 f^2$. Figure 9.15 shows the argand diagram of the steady-state response of $y(t)$ for a particular value of ξ and f. Therefore

$$\phi = -\arctan\left(\frac{2\xi f}{1 - f^2}\right) \tag{9.40}$$

$$= \text{phase lag of response with respect to the harmonic forcing}$$

and

$$Y = \sqrt{(Y_r^2 + Y_i^2)} = \frac{Y_0}{\sqrt{[(1 - f^2)^2 + 4\xi^2 f^2]}} \tag{9.41}$$

$$= \text{amplitude of steady-state response}$$

The response of the system in terms of equations 9.40 and 9.41 is illustrated in graphical form in Figures 9.16a and 9.16b. From these it will be observed that at low values of excitation frequency the amplitude of the dynamic response, Y, differs only marginally from the static or zero frequency response, i.e.

$$Y/Y_0 \approx 1 \qquad \text{for } \Omega \ll \omega$$

When the forcing frequency is much greater than the undamped natural frequency, the vibratory amplitude of the system tends to zero irrespective of the degree of

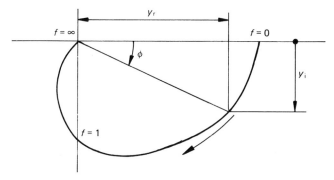

Figure 9.15

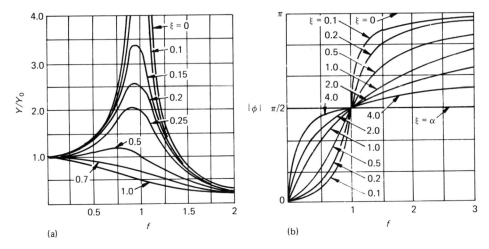

Figure 9.16

damping present. As Ω approaches ω, the response of the system increases dramatically and for the particular case where there is zero damping (i.e. $\xi = 0$) the response is, theoretically, infinite. The condition where the impressed frequency of excitation is exactly equal to the undamped natural frequency of the mass/elastic system is known as **resonance**, a term which is often misused as describing the condition at which the vibratory amplitude reaches a maximum. This is not the case, as may be seen from Figure 9.16a, where the position for maximum amplitude occurs at a frequency lower than that corresponding to the undamped natural frequency of the system ($f = 1$); however, where the degree of damping present is low, the error incurred in such assumptions is small.

If we examine the phase response we see that the condition of resonance, $f = 1$, always corresponds to a phase angle $|\phi| = \pi/2$, *irrespective of the level of damping*. For the special case where no damping is present, phase angle ϕ is 0 for $f < 1$, and π for $f > 1$; in other words the force and displacement are always in phase until resonance is attained, and thereafter they are antiphase to each other. At resonance the phase angle for zero damping is seen to be indeterminate. However, in all practical systems, some degree of damping will always be present and we see that a more realistic value of phase angle, ϕ, is a value ranging between 0 and $\pi/2$ for $f < 1$ and between $\pi/2$ and π for $f > 1$, depending on the level of damping. For any finite degree of damping the condition of resonance always corresponds to a phase angle of $\phi = \pi/2$ exactly. Also, by substituting $f = 1$ in equation 9.41, we have, at resonance,

$$\frac{Y}{Y_0} = \frac{1}{2\xi} \tag{9.42}$$

The degree of damping also controls the frequency of excitation at which the vibratory response reaches a maximum value. This may be demonstrated by differentiating equation 9.41 with respect to the frequency ratio, f, and equating to zero; the resulting expression will be

$$f_{Y_{max}} = \sqrt{(1 - 2\xi^2)} \tag{9.43}$$

Example 9.2

Figure 9.17a represents part of a servomechanism. The oscillation of the rotor about the pivot O is controlled by two springs attached at A, each spring having a linear stiffness of 1.23 N/mm. When given a displacement and released, the rotor oscillates at a frequency of 10 Hz and, after five complete oscillations, the amplitude is reduced to half the initial displacement.

If a harmonic input torque $T(t) = \hat{T} \cos \Omega t$ is applied to the rotor, about point O, where $\hat{T} = 0.14$ N m and $\Omega = 13$ Hz, determine

(a)   the maximum displacement at B,
(b)   the time interval between maximum torque and maximum displacement.

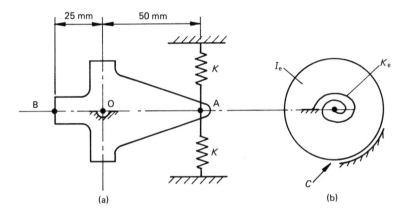

(a)   (b)

Figure 9.17

SOLUTION

(a)   For convenience of solution we shall refer to the system as a simple dynamically equivalent angular system, pivoting about axis O, having equivalent moment of inertia I_e and acted upon by applied torque $T(t)$, elastic stiffness K_e and damping coefficient C, as shown in Figure 9.17b.

From standard dynamic equivalence theory it may be shown that $K_e = 2(K \times \overline{\text{OA}}^2)$ since the springs are acting in parallel. Thus

$$K_e = 2 \times 1.23 \times 10^3 \times (50 \times 10^{-3})^2$$
$$= 6.15 \text{ N m/rad}$$

The basic equation of angular motion of the system is given by

$$I_e \ddot{\theta}(t) + C \dot{\theta}(t) + K_e \theta(t) = \hat{T} \cos \Omega t$$

where $\theta(t)$ is the angular displacement of the rotor from the equilibrium position.

In the absence of the harmonic torque, the subsequent transient motion enables some of the parameters of the system to be determined. First, the damped natural frequency, ω_d, is given as $\omega_d = 10 \times 2\pi = 62.83$ rad/s. Also from equation 9.30 we have the relationship (re-expressed in angular terms)

$$\theta_1/\theta_6 = \exp(5 \times 2\pi\Delta/\omega_d)$$

from which, since $\theta_1/\theta_6 = 1/0.5 = 2$, we obtain

$$\Delta = 1.387 \text{ rad/s}$$

In addition, from the relationship, $\omega_d^2 = \omega^2 - \Delta^2$, we obtain after substitution $\omega = 62.84$ rad/s.

Subsequently since $\omega = \sqrt{(K_e/I_e)}$, we calculate $I_e = I_0 = 1.557 \times 10^{-3}$ kg m². The viscous damping coefficient C may also be determined from $C = 2\Delta I_0 = 4.32 \times 10^{-3}$ N m s/rad. The steady-state solution is

$$\theta(t) = \hat{\theta}\cos(\Omega t + \phi) \qquad \text{where } \Omega = 13 \times 2\pi = 81.68 \text{ rad/s}$$

and

$$\frac{\hat{\theta}}{\hat{\theta}_0} = \frac{1}{\sqrt{[(1-f^2)^2 + 4\xi^2 f^2]}} \tag{1}$$

where

$$\hat{\theta}_0 = \hat{T}/K_e = 0.14/6.15 = 0.0277 \text{ rad}$$
$$f = \Omega/\omega = 81.68/62.84 = 1.3$$
$$\xi = \Delta/\omega = 1.387/62.84 = 0.022$$

Substitution in equation 1 gives

$$\hat{\theta} = 0.033 \text{ rad}$$

The corresponding linear displacement at B, δ_B, will be given by $\overline{OB}\,\hat{\theta}$ (for small angular displacements), i.e.

$$\delta_B = 0.033 \times 25 = 0.82 \text{ mm}$$

(b) The time interval, δt, between the maximum torque and displacement vectors will be given by

$$\delta t = \phi/\Omega$$

where ϕ is the magnitude of the phase angle,

$$\phi = -\arctan\left(\frac{2\xi f}{1-f^2}\right)$$
$$= -175.3 = -3.059 \text{ rad}$$

Therefore $\delta t = 3.059/81.68 = 0.038$ s.
The graphical representation of this solution is shown in Figure 9.18.

9.5.2 Rotational out-of-balance

In rotating machinery the primary source of vibratory excitation is often the out-of-balance of one or more rotating components; in other words, the centre of mass of the rotating system does not correspond to the geometric centre of the supporting bearings through which the axis of rotation passes. If the mass of the rotating system is m and the distance between the axis of rotation, O, and the centre

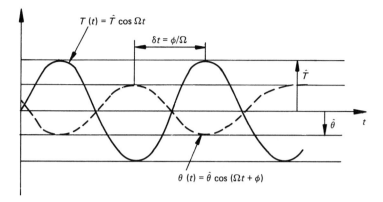

Figure 9.18

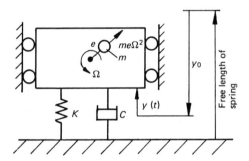

Figure 9.19

of mass for the rotating system (i.e. the eccentricity) is e (as shown in Figure 9.19), then for a given rotational velocity, Ω rad/s, there will be a corresponding rotating force of magnitude equal to $me\Omega^2$ applied to the shaft; the line of action of this force (which is of course the centrifugal force) will be from the axis of rotation radially outwards through the centre of mass of the rotating system, as shown.

It follows, therefore, that the force $p(t)$ examined in Section 9.5.1 will now be replaced by a corresponding force $p(t) = P \cos \Omega t$ where $P = me\Omega^2$. If the total mass of the system, including the rotational mass, is M then equation 9.41 may be re-expressed, remembering that $Y_0 = P/K = me\Omega^2/K$, as

$$\frac{YM}{me} = \frac{f^2}{\sqrt{[(1-f^2)^2 + 4\xi^2 f^2]}} \tag{9.44}$$

which is plotted in Figure 9.20.

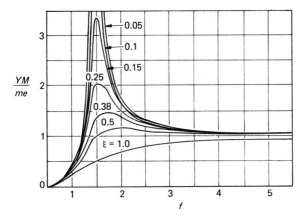

Figure 9.20

The corresponding version of the phase angle is

$$\phi = -\arctan\left(\frac{2\xi f}{1-f^2}\right)$$

which is the same as before. From Figure 9.20 it is evident that, irrespective of the amount of damping present within the system, at rotational speeds greatly in excess of the natural undamped frequency, the steady-state amplitude of vibration of the system tends to a constant value, namely me/M. Consequently, this amplitude can be kept at a very low level by ensuring a large value for M, if reduction of the eccentricity is impossible. This is often achieved in practice by attaching to the machine that is subject to the vibratory excitation an auxiliary mass, which in some cases may be of the same order or even greater than that of the machine; the combined mass system may then be supported on relatively soft springs which in turn will lower the natural frequency of the complete system.

If the response equation 9.44 is differentiated with respect to f and equated to zero we obtain once again the frequency ratio corresponding to the maximum amplitude condition, which in this case is given by

$$f_{Y_{max}} = \frac{1}{(1-2\xi^2)} \tag{9.45}$$

From this relationship and Figure 9.20 it will be seen that for a given degree of damping the frequency corresponding to the maximum amplitude condition is always *greater* than the resonance frequency. However, when the level of damping is low the difference between resonance and maximum amplitude frequencies is negligible.

9.5.3 Foundation force and transmissibility

In the previous section we analysed the motion of a single mass to the response of a harmonic excitation; the subsequent vibratory motion of the mass will in turn, through the mechanisms of elasticity and damping, transfer some if not all of the original force to the supporting structure.

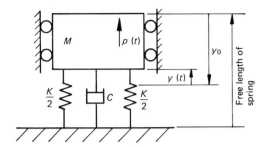

Figure 9.21

This transferred force is referred to as the foundation force, or frame force, and from a vibration, noise and structural integrity aspect should be minimized as far as is practical.

Consider once again the single degree of freedom arrangement shown in Figure 9.21, where the body of mass M is acted upon by a harmonic force $p(t) = P \cos \Omega t$. The harmonic force, $ff(t)$, transmitted to the foundation upon which the system is supported, will be the summation of the elastic and damping forces, namely

$$ff(t) = K\,y(t) + C\,\dot{y}(t) \tag{9.46}$$

Alternatively, from the basic equation of motion, equation 9.33, we have

$$K\,y(t) + C\,\dot{y}(t) = P \cos \Omega t - M\,\ddot{y}(t) \tag{9.47}$$

Therefore, from equations 9.46 and 9.47, we can re-express the foundation (or frame) force, $ff(t)$, as the sum of the applied force and the inertia force acting on the system mass, i.e.

$$ff(t) = P \cos \Omega t + \Omega^2 M Y \cos(\Omega t + \phi) \tag{9.48}$$

As before, we can write

$$ff(t) = (FF_r + i\,FF_i) \cos \Omega t$$

where FF_r and FF_i are the real and imaginary components respectively of the amplitude of $ff(t)$, FF, and i is the complex conjugate, $\sqrt{-1}$. Also from equation 9.36 we have

$$y(t) = (Y_r + i\,Y_i) \cos \Omega t$$

where

$$Y_r = \frac{Y_0(1 - f^2)}{D} \qquad Y_i = \frac{-2\xi f Y_0}{D}$$

and

$$D = (1 - f^2)^2 + 4\xi^2 f^2 \qquad Y_0 = P/K$$

Therefore, substituting for $ff(t)$ and $y(t)$ in equation 9.46, equating real to real and

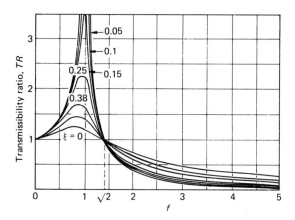

Figure 9.22

imaginary to imaginary terms, gives

$$TR_r = \frac{FF_r}{P} = \frac{(1-f^2)+4\xi^2 f^2}{D}$$

$$TR_i = \frac{FF_i}{P} = \frac{-2\xi f^3}{D}$$

where TR_r and TR_i are the real and imaginary component of the **transmissibility**, TR, such that

$$TR = \sqrt{(TR_r^2 + TR_i^2)}$$

which after substitution and simplification gives

$$TR = FF/P = \frac{\sqrt{(1+4\xi^2 f^2)}}{[(1-f^2)^2 + 4\xi^2 f^2]} \tag{9.49}$$

This ratio, TR, is plotted for various levels of damping in Figure 9.22, where we see that transmissibility is less than unity only when $f > \sqrt{2}$, i.e. when the forcing frequency $\Omega > \sqrt{2}\omega$. In addition the presence of damping reduces the maximum value of the force transmitted to the foundation only when $f < \sqrt{2}$; above this value, damping actually increases the force transmitted. Beyond the point, $f = \sqrt{2}$, **vibration isolation** is said to occur, although this is actually a misleading description since the vibratory force is still present in the structure, albeit at a level below that of the excitation force.

In order to achieve maximum vibration isolation, i.e. minimum transmissibility, the undamped natural frequency of the mass on its support must be made as low as possible either by the use of soft coil springs or by additional mass (as previously mentioned) or by a combination of both.

9.5.4 Seismic excitation and seismic instruments

Seismic excitation

The foundation force of the previous section, although the end product in a chain of energy transmission from the periodic excitation which was applied directly to the single mass, can however be itself a source of subsequent excitation for some other system in close proximity. For example, if the foundation is a factory or office floor, such a floor when vibrating may set into sympathetic vibratory motion some other machine or instrument, which is standing on, or is attached to, the same floor and which forms a separate mass/elastic system.

Vibrations induced by movement of the foundation of a system are described as **seismic** vibrations. Seismic motion may take one of many forms, such as shock loading in the form of a step or ramp input over a very short interval of time or random in nature over a longer timescale.

For the purposes of the present analysis we shall confine the seismic excitation to that of harmonic motion. Consider the mass shown in Figure 9.23 which is subject to a harmonic disturbance, $y(t)$, resulting from a seismic motion, $x(t)$, of the foundation. Elastic and damping forces will, in this case, be proportional, respectively, to the relative displacement and velocity between the mass and the foundation. Applying Newton's Second Law to the body once again, and assuming that at the instant shown $x(t) > y(t)$, gives the equation of motion from the static equilibrium position as

$$K[x(t) - y(t)] + C[\dot{x}(t) - \dot{y}(t)] = M\,\ddot{y}(t) \tag{9.50a}$$

or

$$M\,\ddot{y}(t) + C\,\dot{y}(t) + K\,y(t) = C\,\dot{x}(t) + K\,x(t)$$

or

$$y(t) + 2\xi\omega\,\dot{y}(t) + \omega^2 y(t) = 2\xi\omega\,\dot{x}(t) + \omega^2 x(t) \tag{9.50b}$$

Now, in the usual manner, if $x(t) = X\cos\Omega t$, we can describe $y(t)$ in the form $y(t) = (Y_r + i\,Y_i)\cos\Omega t$, where Y_r and Y_i are the real and imaginary components of the response respectively and i is the complex conjugate, $\sqrt{-1}$. Therefore, substituting for $x(t)$ and $y(t)$ in equation 9.50b, remembering that $\sin\Omega t = (-i)\cos\Omega t$, and equating real to real and imaginary to imaginary terms, gives

$$(1 - f^2)Y_r - 2\xi f Y_i = X \tag{9.51}$$

$$(1 - f^2)Y_i + 2\xi f Y_r = 2\xi f X$$

where $f = \Omega/\omega$, as before. Solving equations 9.51 gives

$$Y_r = \frac{[(1 - f^2) + 4\xi^2 f^2]X}{D} \qquad Y_i = \frac{-2\xi f^3 X}{D}$$

where $D = (1 - f^2)^2 + 4\xi^2 f^2$.

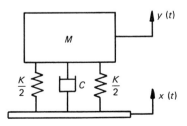

Figure 9.23

Therefore, $Y =$ amplitude of $y(t) = \sqrt{(Y_r^2 + Y_i^2)}$ which after simplification gives

$$\frac{Y}{X} = \frac{\sqrt{(1 + 4\xi^2 f^2)}}{\sqrt{[(1 - f^2)^2 + 4\xi^2 f^2]}} \tag{9.52}$$

The reader will observe that equation 9.52 is identical in form to equation 9.49. Correspondingly the graphical presentation of equation 9.49 in Figure 9.22 will also represent the seismic response of the present system, when the vertical ordinate is changed from transmissibility ratio to Y/X.

The modified form of Figure 9.22 now indicates that at excitation frequencies far beyond the natural undamped frequency of the mass/elastic system (i.e. well above the resonance condition) the displacement of the mass tends to zero. This phenomenon is of fundamental importance to the design and operation of seismic measuring instruments, such as seismographs, vibrographs and torsiographs, since, for frequencies such that $f \gg 1$, the mass can be considered as an artificial datum and the motion of the support, relative to the mass, i.e. $x(t) - y(t)$, may then be considered as the *absolute* motion of the support.

Seismic instruments

Figure 9.24 shows the most common form of a seismic instrument, used for measuring the amplitude and frequency of the foundation movement, $x(t)$. As shown, the position of the pointer (attached to the seismic mass) on the scale (attached to the body of the instrument) is a measurement of the instantaneous position of the seismic mass relative to that of the foundation, which we shall denote as $z(t)$, i.e.

$$z(t) = y(t) - x(t) \tag{9.53}$$

Substituting equation 9.53 in equation 9.50a and rearranging gives

$$M \ddot{z}(t) + C \dot{z}(t) + K z(t) = - M \ddot{x}(t)$$

and, dividing throughout by M,

$$z(t) + 2\xi\omega \dot{z}(t) + \omega^2 z(t) = - \ddot{x}(t) \tag{9.54}$$

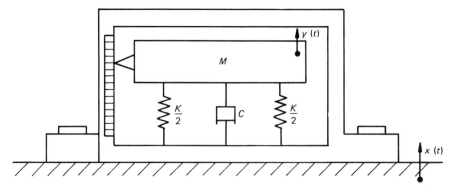

Figure 9.24

As before, if

$$x(t) = X \cos \Omega t \qquad \text{and} \qquad z(t) = (Z_r + i Z_i) \cos \Omega t$$

then after substituting in equation 9.54 and dividing throughout by ω^2 we have

$$-f^2(Z_r + i Z_i) + 2\xi f(-Z_i + i Z_r) + (Z_r + i Z_i) = f^2 X$$

where $f = \Omega/\omega$, as before. Equating real to real and imaginary to imaginary terms results in

$$Z_r = \frac{f^2(1 - f^2)X}{D} \qquad Z_i = \frac{-2\xi f^3 X}{D}$$

where, as before, $D = (1 - f^2)^2 + 4\xi^2 f^2$.
Therefore

$$Z = \sqrt{(Z_r^2 + Z_i^2)} = \frac{f^2 X}{\sqrt{[(1 - f^2)^2 + 4\xi^2 f^2]}} \tag{9.55}$$

where Z is the amplitude of $z(t)(= y(t) - x(t))$.

The term Z/X is known as the **correction factor** for the instrument and, as can be seen from the graphical plot shown in Figure 9.25, tends to unity for high values of f. For example, if the seismic mass is 0.5 critically damped (i.e. $\xi = 0.5$), then for a seismic motion which is, say, at a frequency five times that of the undamped natural frequency of the instrument ($f = 5$), the relevant correction factor (from equation 9.55) would be $+ 1.02$, i.e. the instrument registers a seismic displacement of amplitude some 2 per cent greater than the amplitude of the foundation displacement.

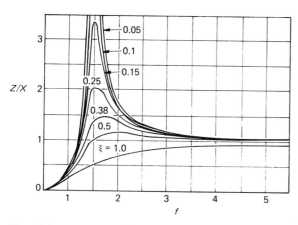

Figure 9.25

Problems

9.1   In a vehicle research laboratory a particular test procedure to assess the strength of car seat-belts consists of driving a car into pneumatic buffers. The buffers have a total equivalent stiffness of 38.5 N/mm and are set to give critical viscous damping, to avoid rebound.

(a) If a car weighing 8.5 kN is driven at the buffers at a speed of 50 km/h, sketch the displacement versus time graph for the system after impact and determine:
    (i) the maximum compression of the buffers,
    (ii) the initial decelerating force sustained by the car on impact with the buffers.
(b) At what speed would the car require to be driven to subject test models and seat-belts within the car to an impact acceleration of 20 g, where g is equal to gravitational acceleration?

Assume that the buffers have negligible mass.

Answer: (a) (i) 0.766 m; (ii) 166.45 k N; (b) 54 km/h.

9.2 In the system shown in Figure 9.26 the rod AD has a mass of 6.3 kg attached at D.

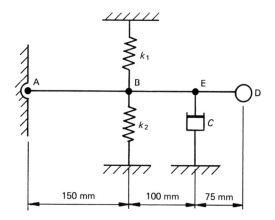

Figure 9.26

The springs k_1 and k_2 at B have stiffnesses of 875 N/m and 350 N/m respectively. Determine:

(a) the natural frequency of the system and the viscous damping coefficient, C, of the dashpot at E if the damping is 0.25 critical,
(b) the magnitude of the applied force at D if the amplitude of vibration at D is restricted to 5 mm at a frequency of 5 Hz,
(c) the phase angle between force and displacement.

Answer: (a) $\omega = 1.024$ Hz, $C = 34.26$ N s/m; (b) 29.96 N; (c) 174°.

9.3 A machine of mass 1100 kg has rotating out-of-balance parts of mass 2 kg at a radius of 150 mm. A static test shows that the weight of the machine causes the mounting springs to deflect by 2.25 mm and a transient test indicates that the amplitude of a free vibration is reduced by 84 per cent over two cycles. Calculate the magnitude of the dynamic force transmitted to the foundation when the machine runs at twice its resonant speed.

Answer: 1.976 kN.

9.4   The vibration transmitted to a floor is troublesome when an engine weighing 2891 N and running at 1500 rev/min is isolated from the floor by cork matting. The weight of the engine compresses the cork by 0.17 mm and a transient test indicates that the amplitude of vibration of the engine is reduced by 80 per cent per cycle of vibration.

It is proposed to replace the cork matting by fitting a spring suspension to the engine of total stiffness 66.34 kN/m. No damping is included in the suspension. The only important out-of-balance force from the engine is given by $P = 0.11\Omega^2 \cos 2\Omega t$ N where Ω is the engine speed in rad/s.

Calculate and compare the dynamic forces transmitted to the floor in both cases.

Answer: 3.42 kN; 6.2 N.

9.5   Figure 9.27 shows a rotor of moment of inertia 0.0124 kg m² driven by an electric motor through a flexible coupling of elastic torsional stiffness 34 N m/rad. At the

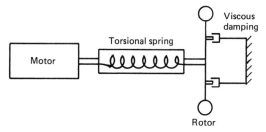

Figure 9.27

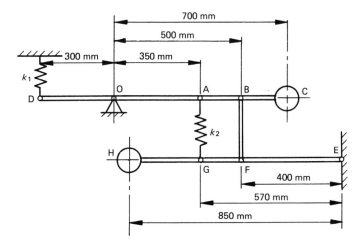

Figure 9.28

operation speed of 750 rev/min, the power transmitted to the rotor is 745 W and it may be assumed that all this power is absorbed by the viscous damping which acts at the rotor.

Calculate the maximum transient vibratory torque that would be transmitted through the couping if the electric motor suddenly stopped due to bearing seizure.

Answer: 43.5 N m.

9.6   Figure 9.28 represents, diagrammatically, part of the mechanism of a balancing machine. The lever DOC weighs 50 N and has a radius of gyration of 100 mm about the pivot O. In addition, a body of weight 5.5 N is attached to the lever at C. The lever EH weight 35 N, has a radius of gyration of 75 mm about the pivot E and has a body of weight 7.5 N attached at H. The springs k_1 and k_2 are of negligible mass and have linear stiffness of 10 N/mm and 6 N/mm respectively.

Determine the natural frequency of angular oscillation of the system, for small angular displacements.

Answer: 5.92 Hz.

10

Free undamped vibration of a two degree of freedom system

10.1 Introduction

In Chapter 9, attention was focussed on the vibration of single degree of freedom systems. However, although it is often possible to reduce fairly complex systems to a simple single mass/elastic equivalent form, there are many situations in which this is no longer the case. In such circumstances a single coordinate may not be sufficient to describe the vibratory motion of the component parts of the system and recourse has to be made to increasing the number of coordinates. Such systems are known as multi degree of freedom systems. As the description implies, these systems consist of many masses, or, in the case of angular systems, rotors, coupled together by elastic components, with the result that the overall arrangement possesses many degrees of freedom, each with a corresponding natural frequency of vibration. A multi-stage axial flow compressor is a typical example of a practical system that possesses many natural frequencies of torsional oscillation, each capable of being excited into resonance by the application of a periodic torque of similar frequency. The detailed analysis of such systems will be the subject of Chapter 11; however, the basis of such analysis can be illustrated by examination of the free undamped vibration of a two mass or rotor system having two degrees of freedom.

10.2 Rectilinear systems

Consider the system shown in Figure 10.1a which consists of two masses M_1 and M_2 connected to each other and to a fixed structure by springs of stiffness K_1, K_2 and K_3 as shown. The masses are constrained to move only in the horizontal direction and it is assumed that damping is negligible. At any given instant let the corresponding displacements of M_1 and M_2, relative to the static equilibrium position, be $x_1 = x_1(t)$ and $x_2 = x_2(t)$, respectively. The free body diagrams of each of the two masses are shown in Figure 10.1b. Therefore applying Newton's Second Law to each of the two masses, and rearranging, gives the equations

$$M_1\ddot{x}_1 + (K_1 + K_2)x_1 - K_2 x_2 = 0 \qquad (10.1)$$

and

$$M_2\ddot{x}_2 - K_2 x_1 + (K_2 + K_3)x_2 = 0$$

Thus the motion of the masses is represented by two simultaneous linear

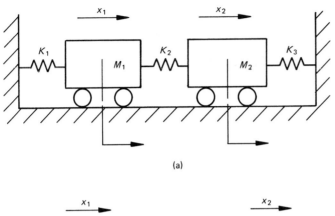

(a)

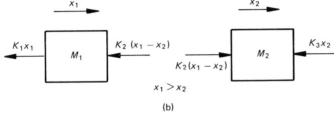

$x_1 > x_2$

(b)

Figure 10.1

homogeneous differential equations having constant coefficients, and may be written in matrix form as

$$\begin{bmatrix} M_1 & 0 \\ 0 & M_2 \end{bmatrix} \begin{Bmatrix} \ddot{x}_1 \\ \ddot{x}_2 \end{Bmatrix} + \begin{bmatrix} (K_1 + K_2) & -K_2 \\ -K_2 & (K_2 + K_3) \end{bmatrix} \begin{Bmatrix} x_1 \\ x_2 \end{Bmatrix} = \begin{Bmatrix} 0 \\ 0 \end{Bmatrix} \tag{10.2}$$

Following the procedure adopted in Chapter 9, let us assume that the system is vibrating in one of its normal or principal modes at a natural frequency, ω, and therefore the displacement of each mass may be written as

$$x_1 = X_1 \cos(\omega t + \phi_1) \tag{10.3}$$

and

$$x_2 = X_2 \cos(\omega t + \phi_2)$$

where X_1 and X_2 are the displacement amplitudes of M_1 and M_2 respectively, and ϕ_1 and ϕ_2 are phase angles. Therefore, substituting for x_1 and x_2 in equation 10.2 gives

$$\begin{bmatrix} (K_1 + K_2 - \omega^2 M_1) & -K_2 \\ -K_2 & (K_2 + K_3 - \omega^2 M_2) \end{bmatrix} \begin{Bmatrix} X_1 \\ X_2 \end{Bmatrix} = \begin{Bmatrix} 0 \\ 0 \end{Bmatrix} \tag{10.4}$$

i.e.

$$[K_d]\{X\} = 0$$

where $[K_d]$ is termed the **dynamic stiffness matrix**, and $\{X\}$ the **response vector**. Therefore, for the case of free vibration, the non-trivial solution to equation 10.4

corresponds to the determinant of $\{X\}$ being equal to zero, which gives

$$M_1 M_2 \omega^4 - [M_1(K_2 + K_3) + M_2(K_1 + K_2)]\omega^2 + [K_1 K_2 + K_1 K_3 + K_2 K_3] = 0 \tag{10.5}$$

Equation 10.5 represents the characteristic equation for the system described, and is a quadratic in ω^2. The two roots of ω^2 that satisfy equation 10.5 are the **latent roots** or **eigenvalues** of the system, ω_1^2 and ω_2^2, such that

$$\omega_1^2 = \frac{-b - \sqrt{(b^2 - 4ac)}}{2a} \quad \text{and} \quad \omega_2 = \frac{-b + \sqrt{(b^2 - 4ac)}}{2a} \tag{10.6}$$

where

$$a = M_1 M_2$$
$$b = -[M_1(K_2 + K_3) + M_2(K_1 + K_2)] \tag{10.7}$$
$$c = (K_1 K_2 + K_1 K_3 + K_2 K_3)$$

By inspection it will be noted that b^2 will always be greater than $4ac$, which implies that both roots are real. Also, since b is negative, $(b^2 - 4ac) < b^2$, signifying that both roots are positive. Therefore, for such a system with known parameters of mass and stiffness, the two (undamped) natural frequencies may be determined. Although it is not possible to obtain the absolute values of X_1 and X_2, the ratio of these maximum displacements may be specified by substitution of ω_1 and ω_2, written as $\omega_j (j = 1, 2)$, into equation 10.4. Therefore

$$(K_1 + K_2 - \omega_j^2 M_1)X_1 - K_2 X_2 = 0 \tag{10.8}$$

and

$$(K_2 + K_3 - \omega_j^2 M_2)X_2 - K_2 X_1 = 0$$

giving

$$\frac{X_1}{X_2} = \frac{K_2}{K_1 + K_2 - \omega_j^2 M_1} \tag{10.9}$$

or

$$\frac{X_1}{X_2} = \frac{K_2 + K_3 - \omega_j^2 M_2}{K_2}$$

The amplitude ratios enable the mode, or shape, of the vibratory deformation of the system to be drawn, and the position of the node or nodes (i.e. positions of zero vibratory displacement) to be determined.

10.3   Torsional systems

Many vibration problems, particularly in the field of power generation and transmission, relate to torsional systems. In these situations the problem manifests itself in the form of angular vibrations which, in cases where the angular displacements are large, can induce large cyclic shear stresses in the connecting shafts. It is important, therefore, that the corresponding analysis of such systems be carried out towards predicting the undamped natural frequencies of the system in order to check that they do not coincide with, and are not close to, any of the frequencies of the torques acting on the system.

10.3.1 General analysis

Figure 10.2a shows the torsional equivalent of the two-mass linear system of Figure 10.1a. The rotors are assumed to move in an angular sense only where θ_1 and θ_2 are the instantaneous angular displacements of the two rotors relative to the static equilibrium position as shown. Therefore the corresponding elastic torques acting on each rotor are shown in Figure 10.2b.

Therefore the equations of motion for the rotors, after rearranging, are

$$I_1\ddot{\theta}_1 + (K_1 + K_2)\theta_1 - K_2\theta_2 = 0$$
$$I_2\ddot{\theta}_2 - K_2\theta_1 + (K_2 + K_3)\theta_2 = 0$$

(10.10)

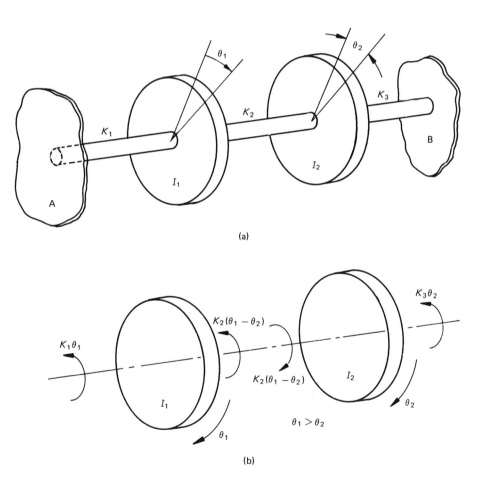

Figure 10.2

These equations are identical in form to equations 10.1. Consequently the solution of equations 10.10, when simple harmonic motion at frequency ω is assumed, will be given by equation 10.5, with the masses M_1 and M_2 being replaced by I_1 and I_2 respectively, and the linear stiffnesses by their torsional counterparts. Therefore, the two undamped natural frequencies can be calculated as before in the linear case, together with the amplitude ratios of the angular displacements at I_1 and I_2 for each of the two resonant states.

10.3.2 Two-rotor single-stiffness systems

Many two-rotor systems come in the form of two rotors connected by a single shaft, e.g. an internal combustion engine driving a flywheel or a multi-stage compressor/turbine combination. Consider, therefore, the two-rotor system shown in Figure 10.3, where the rotors of polar mass moments of inertia I_1 and I_2 are connected by a single shaft of uniform diameter d and length l having an angular stiffness K.

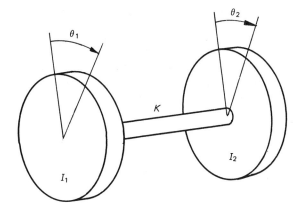

Figure 10.3

If we restrict the analysis once again to the free undamped angular oscillation of the system, the corresponding equations of motion for the rotors may be obtained from equation 10.10 with K_1 and K_3 being zero and $K_2 = K$. Thus we have the following relationships:

$$I_1\ddot{\theta}_1 + K(\theta_1 - \theta_2) = 0 \tag{10.11}$$

and

$$I_2\ddot{\theta}_2 - K(\theta_1 - \theta_2) = 0$$

Assuming, as before, that angular oscillation takes place in the form of simple harmonic motion at an angular frequency, ω, substitution of θ_1, θ_2 and their second derivatives with respect to time in equation 10.11 will give

$$-\omega^2 I_1\hat{\theta}_1 + K(\hat{\theta}_1 - \hat{\theta}_2) = 0 \tag{10.12}$$

and

$$-\omega^2 I_2\hat{\theta}_2 - K(\hat{\theta}_1 - \hat{\theta}_2) = 0 \tag{10.13}$$

where $\hat{\theta}_1$ and $\hat{\theta}_2$ correspond to the maximum angular displacements of rotors I_1 and I_2 respectively. Adding equations 10.12 and 10.13 gives

$$-\omega^2(I_1\hat{\theta}_1 + I_2\hat{\theta}_2) = 0 \tag{10.14}$$

One possible solution to equation 10.14 is $\omega = 0$, which corresponds to the 'zero frequency' or 'rigid body' mode of vibration of the mass/elastic system. In this mode, the combined rotor/shaft system simply rotates about the polar axis of the shaft with no relative angular displacement occurring between the rotors, i.e. angular oscillation does not take place. The other, more important, solution to equation 10.14 is given by

$$I_1\hat{\theta}_1 + I_2\hat{\theta}_2 = 0 \tag{10.15}$$

or

$$\frac{\hat{\theta}_1}{\hat{\theta}_2} = \frac{-I_2}{I_1} \tag{10.16}$$

Equation 10.16 implies that the maximum angular displacements of the rotors are anti-phase to each other and inversely proportional to the respective moments of inertia of the rotors. If either of the peak angular displacements is known, say by measurement, or is assigned a nominal value, then, provided I_1 and I_2 are known, the other peak displacement may be determined from equation 10.16.

If we now substitute for $\hat{\theta}_2 = -\hat{\theta}_1 I_1/I_2$ from equation 10.16 into equation 10.12 we obtain, after further simplification, the frequency equation corresponding to the second principal mode of angular oscillation of the system, namely

$$\omega = \sqrt{\left(\frac{K}{I_1} + \frac{K}{I_2}\right)} = \sqrt{\left(K\frac{I_1 + I_2}{I_1 I_2}\right)} \tag{10.17}$$

10.3.3   Mode of angular oscillation

It will be recalled that equation 10.16 defined the peak amplitude ratio for the angular displacement of the rotors during the principal mode of free angular oscillation of the system. Consequently, this relationship provides two points on a graph relating angular displacement to longitudinal position on the interconnecting shaft. If the diameter of the shaft is considered to be uniform, then from fundamental theory of a uniform bar in torsion we have

$$\theta = (T/GJ_0)l \qquad \text{i.e.} \qquad \theta \propto l \tag{10.18}$$

where T is the applied torque, J_0 is the polar second moment of area of the shaft section, G is the modulus of rigidity of shaft material, and l is the length of shaft over which twist θ rad takes place.

Thus we have a linear relationship between the twist and longitudinal position at any intermediate section of the shaft. Figure 10.4 illustrates the variation of angular displacement at each point along the length of the shaft.

The graph of angular displacement against longitudinal position on the shaft is known as the **normal elastic curve**. This depicts the mode shape or manner of the torsional oscillation. Where the normal elastic curve crosses the axis of the shaft, the angular displacement at that position is zero, i.e. corresponding to a *nodal* position. The position of the nodal points on torsional systems is important from both a measurement and a design consideration, particularly if a gear train forms part of

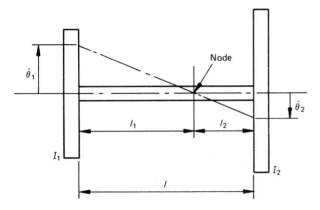

Figure 10.4

the interconnecting link between the rotors. For the system under consideration the position of the node may be determined from similar triangles, i.e.

$$|\hat{\theta}_1/\hat{\theta}_2| = \frac{l_1}{l_2} \qquad (10.19)$$

and from equation 10.16

$$|\hat{\theta}_1/\hat{\theta}_2| = \frac{I_2}{I_1}$$

thus

$$l_1 = \frac{l_2 I_2}{I_1} = \frac{(l - l_1)I_2}{I_1}$$

or

$$l_1 = l\frac{I_2}{I_1 + I_2} \qquad \text{and} \qquad l_2 = l\frac{I_1}{I_1 + I_2} \qquad (10.20)$$

10.3.4 Non-uniform shaft systems

The linear relationship described by equation 10.18 is valid only for the case where the shaft connecting the rotors is of uniform stiffness, i.e. having a constant polar second moment of area, J_0, of the cross-section over the complete length of the shaft. If, as is likely in most practical applications, the shaft is stepped at one or more locations, then, in order to determine the position of the node using equation 10.18, a shaft dynamically equivalent to the actual shaft, but of uniform diameter, must be used. The process of deriving an equivalent shaft has been covered in Chapter 9 and only the end result of the process will be used here. Consider the two rotor system shown in Figure 10.5 where the elastic connection between the rotors I_1 and I_2 is a shaft consisting of three different diameters, i.e. a stepped shaft.

If we refer this system in terms of a dynamically equivalent shaft of uniform diameter d_e, the length of such a shaft, which would exhibit the same angular stiffness as the

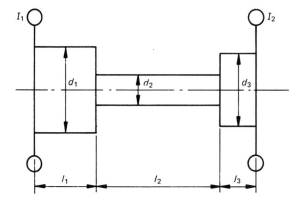

Figure 10.5

original shaft, will be given by the relationship

$$l_e = l_1 \left(\frac{d_e}{d_1}\right)^4 + l_2 \left(\frac{d_e}{d_2}\right)^4 + l_3 \left(\frac{d_e}{d_3}\right)^4 \tag{10.21}$$

assuming that all sections of the shaft are of the same material. The choice of d_e is arbitrary but is usually assigned a value corresponding to one of the existing shaft diameters. The torsional stiffness, K_e, of this equivalent shaft (and of the actual shaft) will therefore be given by $K_e = GJ_e/l_e$, where J_e is the polar second moment of area of the equivalent shaft, and G is the modulus of rigidity of the shaft material, i.e.

$$K_e = \frac{G\pi d_e^4}{32 l_e}$$

and the (non-zero) natural frequency of angular oscillation of the equivalent shaft will be

$$\omega = \sqrt{\left(K_e \frac{I_1 + I_2}{I_1 I_2}\right)} \tag{10.22}$$

The position of the single node associated with this normal mode may be obtained from equations 10.20; this location, however, refers to the nodal point on the *equivalent* shaft. Generally, however, it will be necessary to adopt an inverse equivalent procedure to determine the position of the node on the *actual* shaft. This operation may be more clearly demonstrated by the following example.

Example 10.1

A diesel engine, the rotary parts of which have a polar mass moment of inertia of $I_m = 0.5 \, \text{kg m}^2$, drives a propeller of polar mass moment of inertia $I_p = 5.9 \, \text{kg m}^2$ via a stepped steel propeller shaft as shown in Figure 10.6a. Estimate the natural frequency of the principal mode of angular oscillation of such a system, and determine the position of the nodal point, relative to the engine. Neglect the moment of inertia of the shaft. ($G = 81 \, \text{GN/m}^2$.)

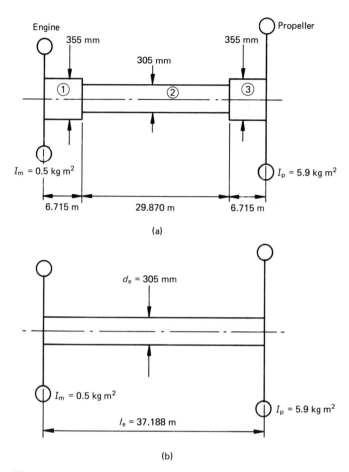

Figure 10.6

SOLUTION

The first step is to derive a dynamically equivalent system consisting of the same rotors but connected by a shaft, having the same angular stiffness as the actual shaft, but having a uniform diameter over the full length of the shaft. For convenience we shall take d_e as 305 mm. Thus the length, l_e, of this equivalent shaft will be

$$l_e = l_1 \left(\frac{d_e}{d_1} \right)^4 + l_2 \left(\frac{d_e}{d_2} \right)^4 + l_3 \left(\frac{d_e}{d_3} \right)^4$$

$$= 2 \times 6.715(305/355)^4 + 29.87 = 37.188 \, \text{m}$$

The dynamically equivalent system is illustrated in Figure 10.6b. The angular stiffness of this equivalent shaft, K_e, is given by

$$K_e = \frac{GJ_e}{l_e}$$

where $J_e = \pi d_e^4/32 = 8.496 \times 10^{-4}\,\text{m}^4$. Thus

$$K_e = 1.851 \times 10^6\,\text{N m/rad}$$

and the natural frequency corresponding to the non-zero principal mode is

$$\omega = \sqrt{\left(K_e \frac{I_m + I_p}{I_m I_p} \right)} = 2004\,\text{rad/s} = 319\,\text{Hz}$$

The position of the node on the equivalent system is given by equation 10.20, which for this particular system may be expressed in the form

$$l_1 = \frac{l_e I_p}{I_m + I_p} = 34.282\,\text{m}$$

i.e. the node is located at a distance of 34.282 m from the engine, on the *equivalent* shaft. This distance consists of two component lengths, namely 3.659 m, which is the equivalent length of the first 355 mm diameter shaft section and $(34.282 - 3.659) = 30.623$ m corresponding to the remaining 305 mm diameter section. It follows therefore that the location of the node on the *actual* shaft will be

$$l_1 = 6.715 + 30.623$$
$$= 37.338\,\text{m from the engine}$$

10.3.5 Torsional vibration of a geared two-rotor system

In the foregoing example the prime mover (i.e. the engine) was connected directly to the load (i.e. the propeller) by means of a single shaft; in other words, both engine and load were running at the same speed. This is a highly inefficient way of transmitting power since the engine will generate optimum power at high speed while the propeller, depending of course on its physical size, will usually represent a high-torque, low-speed energy sink. Greater efficiency will, in most cases, by achieved by the insertion of a gear drive between the power source and the load.

Let two shafts having angular stiffnesses K_1 and K_2, respectively, be geared together as shown in Figure 10.7a. Let each shaft have a rotor attached to the free end having polar mass moments of inertia I_1 and I_2, respectively. Figure 10.7b shows the free body diagrams for all constituent parts of the system where $G_{ab} = \theta_b/\theta_a = r_a/r_b$ where r_a and r_b are the pitch circle radii of gears A and B respectively. F is the tangential action and reaction force acting at the point of contact of the gears A and B. Applying Newton's Second Law to each of the free body diagrams, assuming that each of the two gears have negligible mass, we have

$$
\left.
\begin{aligned}
\text{for } I_1: \quad & -K_1(\theta_1 - \theta_a) = I_1 \frac{d^2\theta_1}{dt^2} \\[2mm]
\text{for A:} \quad & K_1(\theta_1 - \theta_a) - F r_a = 0 \\[2mm]
\text{for B:} \quad & F r_b - K_2(\theta_2 + G_{ab}\theta_a) = 0 \\[2mm]
\text{for } I_2: \quad & -K_2(\theta_2 + G_{ab}\theta_a) = I_2 \frac{d^2\theta_2}{dt^2}
\end{aligned}
\right\} \tag{10.23}
$$

If we now write $\theta_1 = \hat{\theta}_1 \cos \omega t$, $\theta_2 = \hat{\theta}_2 \cos \omega t$, etc., the above equations simplify to

$$\text{for } I_1: \quad (K_1 - \omega^2 I_1)\hat{\theta}_1 - K_1 \hat{\theta}_a = 0 \tag{10.24}$$

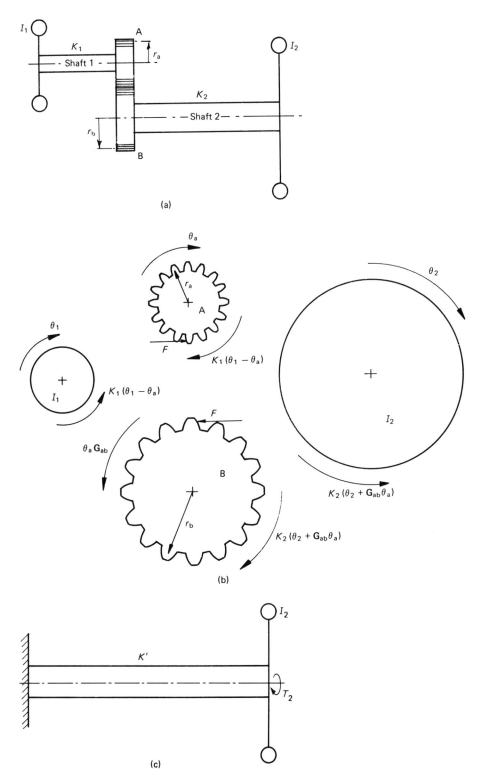

Figure 10.7

for I_2: $(K_2 - \omega^2 I_2)\hat{\theta}_2 + K_2 \mathbf{G}_{ab}\hat{\theta}_a = 0$ (10.25)

for A: $K_1(\hat{\theta}_1 - \hat{\theta}_a) = Fr_a$ (10.26)

for B: $K_2(\hat{\theta}_2 + \mathbf{G}_{ab}\hat{\theta}_a) = Fr_b$ (10.27)

If we now multiply equation 10.27 b6 $\mathbf{G}_{ab} = r_a/r_b$, and then compare the result with equation 10.26, we have

$$K_1(\hat{\theta}_1 - \hat{\theta}_a) = \mathbf{G}_{ab}K_2(\hat{\theta}_2 + \mathbf{G}_{ab}\hat{\theta}_a)$$

i.e.

$$\hat{\theta}_a = \alpha K_1 \hat{\theta}_1 - \alpha \mathbf{G}_{ab}K_2 \hat{\theta}_2 \tag{10.28}$$

where $\alpha = 1/(K_1 + \mathbf{G}_{ab}^2 K_2)$. Therefore substituting for θ_a in equations 10.24 and 10.25 gives the matrix equation

$$\begin{bmatrix} K_1 - \omega^2 I_1 - \alpha K_1^2 & \alpha \mathbf{G}_{ab}K_1 K_2 \\ \alpha K_1 K_2 \mathbf{G}_{ab} & K_2 - \omega^2 I_2 - \alpha \mathbf{G}_{ab}^2 K_2^2 \end{bmatrix} \begin{Bmatrix} \hat{\theta}_1 \\ \hat{\theta}_2 \end{Bmatrix} = \begin{Bmatrix} 0 \\ 0 \end{Bmatrix} \tag{10.29}$$

Equating the determinant of equation 10.29 to zero we find after simplification

$$\omega^2 \{I_1 I_2 \omega^2 - [I_1 K_2(1 - \alpha \mathbf{G}_{ab}^2 K_2) + I_2 K_1(1 - \alpha K_1)]\} = 0$$

which, for the 'non-rolling' mode, is

$$\omega^2 = \frac{I_1 K_2(1 - \alpha \mathbf{G}_{ab}^2 K_2) + I_2 K_1(1 - \alpha K_1)}{I_1 I_2} \tag{10.30}$$

Substitution of ω^2 back into the equation 10.29 will enable one to solve for the normal elastic curve between $\hat{\theta}_1$ and $\hat{\theta}_2$, although it is normal to set $\hat{\theta}_1 = 1$. Subsequently, $\hat{\theta}_a$ can be solved for from equation 10.28.

Alternatively, this type of problem may be handled by constructing a dynamically equivalent single-shaft, single-speed system which has dynamically identical characteristics to the actual physical system; this may be achieved in the following manner.

Referring to Figure 10.7a let one of these rotors, say I_1, be held rigidly (thus reducing it to a static system) whilst a torque is applied to the other, I_2, such that ϕ_1 is the subsequent angle of twist in shaft 1 whilst ϕ_2 is the corresponding angle of twist in shaft 2. It follows, therefore, that if T_1 and T_2 are the respective torques transmitted through shafts 1 and 2 then

$$\phi_1 = \frac{T_1}{K_1} \quad \text{and} \quad \phi_2 = \frac{T_2}{K_2} \tag{10.31}$$

From the geometry of the gear train, we have the relationship

$$\frac{T_2}{T_1} = \mathbf{G}_{ab} = \frac{n_b}{n_a}$$

where n_a and n_b are the number of teeth on the gears A and B respectively. Thus

$$T_1 = \frac{T_2}{\mathbf{G}_{ba}} \tag{10.32}$$

If we refer the two-shaft system as an equivalent single, uniform diameter shaft at the *line of shaft 2* as shown in Figure 10.7c, then the criterion which requires to be satisfied is that the strain energy (SE) induced in both actual and equivalent systems (under the action of T_2) must be the same.

Consequently the total *SE* in the actual two shaft system will be

$$SE_{act} = \tfrac{1}{2}K_1\phi_1^2 + \tfrac{1}{2}K_2\phi_2^2 \tag{10.33}$$

and substituting for ϕ_1 and ϕ_2 gives

$$SE_{act} = \tfrac{1}{2}\frac{T_2^2}{K_1 G_{ba}^2} + \tfrac{1}{2}\frac{T_2^2}{K_2} = \tfrac{1}{2}T_2^2\left(\frac{1}{K_1 G_{ba}^2} + \frac{1}{K_2}\right) \tag{10.34}$$

For the single equivalent shaft the associated strain energy will be given by

$$SE_{eq} = \tfrac{1}{2}K'\phi_e^2$$

where ϕ_e is the total twist in the equivalent shaft, and $\phi_e = T_2/K'$. Thus

$$SE_{eq} = \tfrac{1}{2}\frac{T_2^2}{K'} \tag{10.35}$$

Equating equations 10.34 and 10.35 for dynamical equivalence produces the relationship

$$\frac{1}{K'} = \frac{1}{K_2} + \frac{1}{G_{ba}^2 K_1} \tag{10.36}$$

Referal of shaft stiffnesses, therefore, follows closely the manner of referal of moments of inertia, i.e. by the square of the gear (or speed) ratio between shafts. To demonstrate the effect of a gear train in the torsional oscillation of a transmission system, let us extend Example 10.1 by interposing a 2:1 reduction gear train between the engine and propeller shafts with the engine shaft being 3 m long and of uniform diameter 150 mm. What effect does this have on (a) the natural frequency of the system and (b) the position of the node?

If we refer to the two-shaft system as a dynamically equivalent single, uniform diameter shaft system at, say, the line of the propeller shaft, then the angular stiffness of such a shaft will be given by equation 10.36 with $G_{ba} = 2$. Furthermore, in this case,

$$K_1 = \frac{GJ_1}{l_1} = \text{torsional stiffness of the engine shaft}$$

$$= \frac{81 \times 10^9}{3} \times \frac{\pi \times 0.15^4}{32} = 1.342 \times 10^6 \text{ N m/rad}$$

and K_2 (torsional stiffness of the propeller shaft) will be unchanged at 1.851×10^6 N m/rad. Substituting in equation 10.36 gives

$K' = $ torsional stiffness of a single dynamically equivalent shaft of 350 mm uniform
      diameter
    $= 1.376 \times 10^6$ N m/rad

(Note that the inclusion of this additional shaft has reduced the overall angular stiffness of the system.)

If we now substitute in equation 10.17 the value K' for K, $I' = G_{ba}^2 I_m$ (the dynamically equivalent moment of inertia of the engine when referred to the propeller shaft) for I_1, and I_p for I_2, then

$$\omega = 959 \text{ rad/s} = 152.7 \text{ Hz} \quad \text{(cf. 319 Hz previously)}$$

For an overall equivalent shaft stiffness of 1.376×10^6 N m/rad the equivalent length l' of such a shaft, having a uniform diameter of 305 mm, would be

$$l' = \frac{GJ_e}{K'} = 50.01 \text{ m}$$

The corresponding distance of the node from the engine on the equivalent shaft is given by equation 10.20 thus:

$$l_1 = \frac{l'I_p}{I' + I_p} = 37.35 \text{ m}$$

The question now arises: is the node located on the engine or propeller shaft? To answer this we need to know the relative position of the gear train on the equivalent shaft. It will be recalled that the equivalent stiffness of the (150 mm diameter) engine shaft when referred to the propeller shaft was $G_{ba}^2 K_1 = 2^2 \times 1.342 \times 10^6 = 5.368 \times 10^6$ N m/rad. This now has to be represented in terms of an equivalent length, l_e, of 305 mm diameter shaft of the same material. Thus from the relationship

$$\frac{GJ_e}{l_e} = 5.368 \times 10^6 \text{ N m/rad}$$

we obtain the relevant equivalent length, l_e:

$$l_e = \frac{81 \times 10^9}{5.368 \times 10^6} \times \frac{\pi}{32} \times 0.305^4 = 12.82 \text{ m}$$

It follows, therefore, that the line of the gear train connecting the engine and

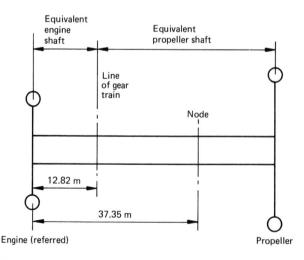

Figure 10.8

propeller shafts is located at a distance of 12.82 mm from the engine end of the equivalent shaft; consequently the node lies in the propeller shaft as shown in Figure 10.8. The corresponding distance between the node and the gear train, on the equivalent shaft, will be $(37.35 - 12.82) = 24.53$ m.

However, this includes 3.654 m of 305 mm diameter shaft which was shown to be equivalent to 6.715 m of 355 mm diameter shaft on the actual system. Thus the position of the node in relation to the gear train, on the *actual* propeller shaft, will be $(24.53 - 3.654) + 6.715 = 27.591$ m.

Problems

In all cases, take $G = 80\,\text{GN/m}^2$.

10.1 The drive between an electric motor and a machine consists of a flexible coupling followed by 608 mm of 75 mm diameter shaft. The moments of inertia of motor and machine are 28.77 and 102.43 kg m² respectively. With the machine clamped stationary, the motor vibrates torsionally at 9.83 Hz.

   Calculate the torsional stiffness of the flexible coupling and the frequency of torsional vibration of the system when the machine is unclamped. For this condition determine the position of the node.

   Answer: 14.993×10^4 N m/rad; 11.14 Hz; 0.500 m.

10.2 A motor drives a machine through a quill drive and a double-reduction gear as shown in Figure 10.9.

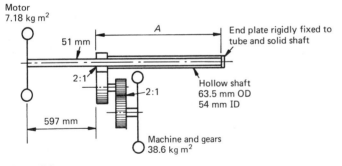

Motor
7.18 kg m²

A

End plate rigidly fixed to tube and solid shaft

51 mm

2:1

2:1

Hollow shaft
63.5 mm OD
54 mm ID

597 mm

Machine and gears
38.6 kg m²

Figure 10.9

   Determine the length A so that the system shall have a natural frequency of torsional oscillation of 1250 cycles/min.

   Calculate and show on a sketch the position of the node when the system oscillates at this frequency.

   Answer: 0.596 m; 0.430 m from motor.

10.3 An engine drives a torque converter via two flexible couplings and a 3:1 step-up gear as shown in Figure 10.10. The shafting may be treated as rigid. Calculate the natural frequency of torsional vibration of the system.

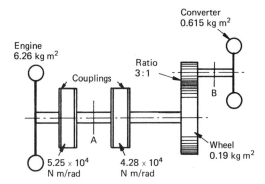

Figure 10.10

If the resonant amplitude of the engine is 1°, estimate the vibratory torque transmitted by the shaft at A and B.

Answer: 848 vibrations/min; 862 and 289 N m.

11

Vibration of multiple degree of freedom lumped mass systems— matrix analysis

11.1 Introduction

In Chapter 10 we confined our study to that of the free undamped vibration of a two degree of freedom system. A large proportion of real systems and structures, however, have multiple degrees of freedom and as such their analysis is often complicated by the correspondingly large number of equations of motion to be solved. With the advent of high-speed/storage digital computers, however, such systems are handled by expressing the equations of motion in terms of matrices, whereupon the computer software implements the relevent matrix manipulative operations. Thus, with the use of such computers, systems with very large numbers of degrees of freedom can be analysed not only to predict the undamped natural frequencies and associated normal mode shapes, but also to predict the forced response of the system in the presence of damping. Finally, as we shall see at a later stage in this chapter, by means of modal analysis it is possible to estimate certain values describing the all-important parameters of mass, stiffness and damping of a system from experimentally obtained measurements of the system's response to known and controlled forcing.

For the purpose of demonstration we will, throughout most of this chapter, apply the principles to the simple two degree of freedom system shown in Figure 11.1 where C_{01}, C_{12} and C_{20} are damping coefficients and k, m, f and d are consistent units of stiffness, mass, force and displacement respectively. Also F_1 and F_2 are time-dependent forcing functions (non-dimensional) acting on the two masses as shown. The reason for adopting this style of presentation is that when one is dealing with matrices and matrix vectors it is always to be recommended that the elements contained within them be dimensionless.

11.2 Undamped natural frequencies and associated normal modes (eigenvalues and eigenvectors)

Consider the system shown in Figure 11.2 in a state of free undamped vibration, i.e. we remove all the forcing and damping coefficients from Figure 11.1. From equation 10.4 of Section 10.2 we have the equation of motion for each mass, expressed in matrix form, as

$$m\begin{bmatrix} 2 & 0 \\ 0 & 3 \end{bmatrix}\begin{Bmatrix} \ddot{x}_1 \\ \ddot{x}_2 \end{Bmatrix}d + k\begin{bmatrix} 5 & -3 \\ -3 & 5 \end{bmatrix}\begin{Bmatrix} x_1 \\ x_2 \end{Bmatrix}d = \begin{Bmatrix} 0 \\ 0 \end{Bmatrix} \tag{11.1}$$

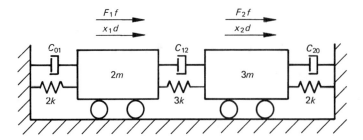

Figure 11.1

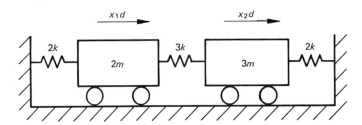

Figure 11.2

and, as before, if $x_j = X_j \cos(\omega t + \phi_j), j = 1, 2$, then

$$\begin{bmatrix} 5 & -3 \\ -3 & 5 \end{bmatrix} \begin{Bmatrix} X_1 \\ X_2 \end{Bmatrix} = \lambda \begin{bmatrix} 2 & 0 \\ 0 & 3 \end{bmatrix} \begin{Bmatrix} X_1 \\ X_2 \end{Bmatrix} \tag{11.2}$$

where $\lambda = \omega^2 m/k$. Equation 11.2 can be written in general terms thus:

$$[K]\{\psi\} = \lambda[M]\{\psi\} \tag{11.3}$$

where $\{\psi\} = \{X_1 \ X_2\}^T$, and $[K]$ and $[M]$ are the non-dimensionalized stiffness and mass matrices of the system respectively *which are always symmetric*. Furthermore, since for the case of lumped mass systems, such as the present one, the mass matrix is a diagonal matrix, equation 11.2 can be written as

$$\begin{bmatrix} \tfrac{1}{2} & 0 \\ 0 & \tfrac{1}{3} \end{bmatrix} \begin{bmatrix} 5 & -3 \\ -3 & 5 \end{bmatrix} \begin{Bmatrix} X_1 \\ X_2 \end{Bmatrix} - \lambda \begin{bmatrix} 1 & 0 \\ 0 & 1 \end{bmatrix} \begin{Bmatrix} X_1 \\ X_2 \end{Bmatrix} = \begin{Bmatrix} 0 \\ 0 \end{Bmatrix}$$

i.e.

$$\begin{bmatrix} \tfrac{5}{2} & -\tfrac{3}{2} \\ -1 & \tfrac{5}{3} \end{bmatrix} \begin{Bmatrix} X_1 \\ X_2 \end{Bmatrix} - \lambda \begin{bmatrix} 1 & 0 \\ 0 & 1 \end{bmatrix} \begin{Bmatrix} X_1 \\ X_2 \end{Bmatrix} = \begin{Bmatrix} 0 \\ 0 \end{Bmatrix} \tag{11.4a}$$

or, in general terms,

$$[A]\{\psi\} - \lambda[I]\{\psi\} = \{0\} \tag{11.5}$$

With reference to equation 11.5, λ and $\{\psi\}$ are termed the **eigenvalue** and associated **eigenvector** respectively of the **dynamic matrix**, $[A]$. Furthermore, as we have already seen in Section 10.2, the eigenvectors have no specific absolute values, but are merely relative values, hence it is convenient to put $X_1 = 1$. Therefore expressing equation 11.4a in the form

$$\begin{bmatrix} (\frac{5}{2} - \lambda) & -\frac{3}{2} \\ -1 & (\frac{5}{3} - \lambda) \end{bmatrix} \begin{Bmatrix} 1 \\ X_2 \end{Bmatrix} = \begin{Bmatrix} 0 \\ 0 \end{Bmatrix} \tag{11.4b}$$

the two values of λ satisfying the above equation may be obtained by equating the determinant of the matrix on the left-hand side to zero, i.e.

$$\lambda^2 - 4.1667\lambda + 2.6667 = 0$$

giving the two roots of $\lambda(= \omega^2 m/k)$ as $\lambda_1 = 0.7897$ and $\lambda_2 = 3.377$. Now substituting λ_1 for λ in equation 11.4b gives $X_2 = 1.1402$. Similarly substituting λ_2 for λ, gives $X_2 = -0.5847$. Therefore it can be said that the system described has eigenvalues of $\lambda_1 = 0.7897$ and $\lambda_2 = 3.377$ with associated eigenvectors $\{\psi_1\} = \{1 \quad 1.1402\}^T$ and $\{\psi_2\} = \{1 \quad -0.5847\}^T$ respectively. Also, if the units k and m are defined, the corresponding two undamped natural frequencies, ω_1 and ω_2, can be calculated from $\omega_j = \sqrt{(\lambda_j k/m)}$, $(j = 1, 2)$. In systems with larger numbers of degrees of freedom, say n, the n eigenvalues and associated eigenvectors are usually computed by means of various numerical iterative techniques. Although description of such techniques is outside the scope of this text, the reader may wish to refer to some of the numerous texts relating to mathematical numerical techniques, which usually contain exhaustive sections dealing with the computation of eigenvalues and eigenvectors, which are sometimes termed **latent roots** and **latent vectors** respectively.

11.2.1   Orthogonal properties of eigenvectors

Consider for the present a general n degree of freedom system having therefore n eigenvalues, λ_j, and associated eigenvectors, $\{\psi_j\}$, $(j = 1, n)$. Consequently for any particular eigenvalue, λ_j, and associated eigenvector, $\{\psi_j\}$, we have from equation 11.3

$$[K]\{\psi_j\} = \lambda_j[M]\{\psi_j\} \tag{11.6}$$

Now, if we premultiply by the transpose of *any other* eigenvector, $\{\psi_i\}$, we have

$$\{\psi_i\}^T[K]\{\psi_j\} = \lambda_j\{\psi_i\}^T[M]\{\psi_j\} \tag{11.7}$$

Similarly if we write down the appropriate form of equation 11.6 for the ith eigenvector and premultiply by the transpose of the jth eigenvector, i.e.

$$\{\psi_j\}^T[K]\{\psi_i\} = \lambda_i\{\psi_j\}^T[M]\{\psi_i\} \tag{11.8}$$

Also, since $[K]$ and $[M]$ are symmetric matrices,

$$\{\psi_j\}^T[K]\{\psi_i\} = \{\psi_i\}^T[K]\{\psi_j\} \tag{11.9}$$

and

$$\{\psi_j\}^T[M]\{\psi_i\} = \{\psi_i\}^T[M]\{\psi_j\}$$

Therefore, subtracting equation 11.8 from equation 11.7, with equation 11.9 implied, gives

$$0 = (\lambda_j - \lambda_i)\{\psi_i\}^T[M]\{\psi_j\} \tag{11.10}$$

and if $\lambda_j \neq \lambda_i$ then

$$\{\psi_i\}^T[M]\{\psi_j\} = 0 \tag{11.11a}$$

and from equation 11.8 we have

$$\{\psi_i\}^T[K]\{\psi_j\} = 0 \tag{11.11b}$$

Equations 11.11a and 11.11b define the orthogonal properties of the eigenvectors with respect to the system mass and stiffness matrices respectively. Furthermore, it can be seen from equation 11.10 that if $i = j$ the products on the left-hand side of equations 11.11a and 11.11b need not be zero, but will in general be equal to some scalar constants, i.e. m_j and k_j respectively, and are termed the **modal mass** and **modal stiffness** values respectively associated with the jth mode of undamped vibration. Also, for the case where $i = j$, from equation 11.7 it can be seen that $\mathbf{k}_j = \lambda_j \mathbf{m}_j$, i.e. $\lambda_j = \mathbf{k}_j/\mathbf{m}_j$.

Let us now apply the above to the two degree of freedom system of Figure 11.2, where

$$[M] = \begin{bmatrix} 2 & 0 \\ 0 & 3 \end{bmatrix} \qquad [K] = \begin{bmatrix} 5 & -3 \\ -3 & 5 \end{bmatrix}$$

and

$$\{\psi_1\} = \{1 \quad 1.1402\}^T$$
$$\{\psi_2\} = \{1 \quad -0.5847\}^T$$

Hence

$$\{\psi_1\}^T[M]\{\psi_1\} = 5.9 = \mathbf{m}_1$$
$$\{\psi_1\}^T[K]\{\psi_1\} = 4.6589 = \mathbf{k}_1$$

and $\lambda_1 = \mathbf{k}_1/\mathbf{m}_1 = 4.6589/5.9 = 0.7897$ (as before).

Also,

$$\{\psi_2\}^T[M]\{\psi_2\} = 3.0256 = \mathbf{m}_2$$
$$\{\psi_2\}^T[K]\{\psi_2\} = 10.2176 = \mathbf{k}_2$$

and $\lambda_2 = \mathbf{k}_2/\mathbf{m}_2 = 10.2176/3.0256 = 3.377$ (as before).

Furthermore, the reader is encouraged to compute the products $\{\psi_1\}^T[K]\{\psi_2\}$, $\{\psi_2\}^T[K]\{\psi_1\}$, $\{\psi_1\}^T[M]\{\psi_2\}$ and $\{\psi_2\}^T[M]\{\psi_1\}$ to prove that, by virtue of the orthogonal properties of eigenvectors, they are all zero.

Let us now proceed to demonstrate how this important property can be used to determine (a) the steady-state response of the systems when subjected to harmonic forcing, and (b) the transient response when given initial displacements and/or velocities.

11.3   Response of undamped and damped systems—modal analysis

11.3.1   Steady-state response to harmonic forcing

Undamped systems

Consider once again our simple two degree of freedom undamped system acted upon by two time-dependent forces F_1 and F_2 as shown in Figure 11.3. If we now express

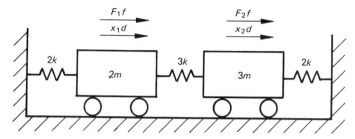

Figure 11.3

the equations of motion for each of the two masses in matrix form, we have

$$m\begin{bmatrix} 2 & 0 \\ 0 & 3 \end{bmatrix}\begin{Bmatrix} \ddot{x}_1 \\ \ddot{x}_2 \end{Bmatrix}d + k\begin{bmatrix} 5 & -3 \\ -3 & 5 \end{bmatrix}\begin{Bmatrix} x_1 \\ x_2 \end{Bmatrix}d = \begin{Bmatrix} F_1 \\ F_2 \end{Bmatrix}f \qquad (11.12)$$

i.e.

$$m[M]\{\ddot{x}\}d + k[K]\{x\}d = \{F\}\,f \qquad (11.13)$$

From equation 11.12 we can deduce that the solution for x_1 and x_2 is complicated by the fact that the matrix $[K]$ is not a diagonal matrix, i.e. the equations of motion describing x_1 and x_2 are **coupled**. Thus, if it were generally possible to apply some transformation such that both matrices, $[K]$ and $[M]$, were made to be diagonal then a solution to equation 11.12 would be simply achieved by sequentially solving the resulting uncoupled, or independent, equations of motion. The process of deriving the system response by transforming the equations of motion into an independent set of equations is known as **modal analysis**. It is here that the orthogonal properties of eigenvectors come into use, since in Section 11.2.1 it was shown that if the stiffness or mass matrix is post- and pre-multiplied by an eigenvector and its transpose, respectively, the result is some scalar constant. Similarly, if either of the two matrices is post-multiplied by an eigenvector and then pre-multiplied by the transpose of *another* eigenvector, the result is zero. Thus with the use of a matrix $[\mathbf{V}]$ the columns of which are the system eigenvectors, we have a transformation of the form

$$\{x\} = [\mathbf{V}]\{\mathbf{q}\} \qquad (11.14)$$

where the matrix $[\mathbf{V}]$ is termed the **modal matrix**, which in the present example is

$$[\mathbf{V}] = \begin{bmatrix} 1 & 1 \\ 1.1402 & -0.5847 \end{bmatrix}$$

and the vector $\{\mathbf{q}\}$ is termed the vector of **modal coordinates**. Substituting equation 11.14 into equation 11.13 and pre-multiplying throughout by $[\mathbf{V}]^{\mathrm{T}}$ gives

$$m[\mathbf{V}]^{\mathrm{T}}[M][\mathbf{V}]\{\ddot{\mathbf{q}}\}d + k[\mathbf{V}]^{\mathrm{T}}[K][\mathbf{V}]\{\mathbf{q}\}d = [\mathbf{V}]^{\mathrm{T}}\{F\}\,f \qquad (11.15)$$

which when applied to equation 11.12, describing the present system, gives

$$m\begin{bmatrix} \mathbf{m}_1 & 0 \\ 0 & \mathbf{m}_2 \end{bmatrix}\begin{Bmatrix} \ddot{\mathbf{q}}_1 \\ \ddot{\mathbf{q}}_2 \end{Bmatrix}d + k\begin{bmatrix} \mathbf{k}_1 & 0 \\ 0 & \mathbf{k}_2 \end{bmatrix}\begin{Bmatrix} \mathbf{q}_1 \\ \mathbf{q}_2 \end{Bmatrix}d = \begin{bmatrix} 1 & 1.1402 \\ 1 & -0.5847 \end{bmatrix}\begin{Bmatrix} F_1 \\ F_2 \end{Bmatrix}f$$

i.e.

$$m[\mathbf{M}]\{\ddot{\mathbf{q}}\}d + k[\mathbf{K}]\{\mathbf{q}\}d = [\mathbf{V}]^{\mathrm{T}}\{F\}\,f \qquad (11.16)$$

where $[\mathbf{M}]$ and $[\mathbf{K}]$ are termed the **modal mass** and **modal stiffness matrices** respectively and, as before, $\mathbf{m}_1 = 5.9$, $\mathbf{m}_2 = 3.1256$, $\mathbf{k}_1 = 4.6589$ and $\mathbf{k}_2 = 10.2176$. Also it was shown in Section 11.2.1 that $\lambda_1 = \mathbf{k}_1/\mathbf{m}_1$ and $\lambda_2 = \mathbf{k}_2/\mathbf{m}_2$. Therefore, from equation 11.16 we can write

$$\ddot{\mathbf{q}}_j + \lambda_j(k/m)\mathbf{q}_j = (1/\mathbf{m}_j)\{\psi_j\}^{\mathrm{T}}\{F\}\, f/md \qquad j = 1, 2 \tag{11.17a}$$

Also, since $\lambda_j = \mathbf{k}_j/\mathbf{m}_j = \omega_j^2 m/k$, equation 11.17a can be expressed in the form

$$\ddot{\mathbf{q}}_j + \omega_j^2\mathbf{q}_j = (\omega_j^2/k)\{\psi_j\}^{\mathrm{T}}\{F\}\, f/kd \qquad j = 1, 2 \tag{11.17b}$$

Hence, if the force vector $\{F\}$ is specified, equation 11.17a or 11.17b can be solved independently for the steady-state response of the modal coordinates, $\mathbf{q}_1$ and $\mathbf{q}_2$. Subsequently, we can solve for the steady-state response of x_1 and x_2 from equation 11.14.

Example 11.1

For the system shown in Figure 11.3, $F_1 = 2\cos 30t$ and $F_2 = 3\sin 60t$. Describe the steady-state response of both masses when $m = 1\,\text{kg}$, $k = 1000\,\text{N/m}$, $f = 1\,\text{N}$ and $d = 1\,\text{m}$.

SOLUTION

From equation 11.17a we have

$$\ddot{\mathbf{q}}_1 + 789.7\mathbf{q}_1 = \frac{2\cos 30t + 3.4206\sin 60t}{5.9}$$

$$\ddot{\mathbf{q}}_2 + 3377\mathbf{q}_2 = \frac{2\cos 30t - 1.7541\sin 60t}{3.0256}$$

and solving gives

$$\mathbf{q}_1 = -(3\cos 30t + 0.206\sin 60t) \times 10^{-3}$$

and

$$\mathbf{q}_2 = (0.27\cos 30t + 2.6\sin 60t) \times 10^{-3}$$

Hence, substituting for $\mathbf{q}_1$ and $\mathbf{q}_2$ in equation 11.14 gives

$$x_1 = -(2.73\cos 30t - 2.394\sin 60t) \times 10^{-3}$$

and

$$x_2 = -(3.578\cos 30t + 1.755\sin 60t) \times 10^{-3}$$

Note that, in this particular example, the forcing frequencies of 30 and 60 rad/s are reasonably close to the undamped natural frequencies $\omega_1(= 28.1\,\text{rad/s})$ and $\omega_2(= 58.11\,\text{rad/s})$ respectively.

Damped systems

In this section we shall consider only the case where the damping is assumed to be *viscous*, i.e. the damping force is proportional to velocity. Furthermore, in certain systems and structure, the viscous damping coefficients can be assumed to be proportional to the corresponding stiffness or mass coefficients, or some combination of both. In other systems, however, the damping coefficients are totally independent of the stiffness or mass coefficients. In this section we shall consider both cases.

(i) *Proportional viscous damping.* Let us consider now the system described in Figure 11.1. Writing down the equations of motion for each of the two masses and expressing them in matrix form gives

$$m\begin{bmatrix} 3 & 0 \\ 0 & 2 \end{bmatrix}\begin{Bmatrix} \ddot{x}_1 \\ \ddot{x}_2 \end{Bmatrix} + \begin{bmatrix} C_{01} + C_{12} & \vdots & -C_{12} \\ \text{----} & \vdots & \text{----} \\ -C_{12} & \vdots & C_{12} + C_{20} \end{bmatrix}\begin{Bmatrix} \dot{x}_1 \\ \dot{x}_2 \end{Bmatrix}$$

$$+ k\begin{bmatrix} 5 & -3 \\ -3 & 5 \end{bmatrix}\begin{Bmatrix} x_1 \\ x_2 \end{Bmatrix} = \begin{Bmatrix} F_1 \\ F_2 \end{Bmatrix} f/d \tag{11.18}$$

i.e.

$$m[M]\{\ddot{x}\} + c[C]\{\dot{x}\} + k[K]\{x\} = \{F\}\, f/d \tag{11.19}$$

where c is a consistent unit of viscous damping and $[C]$ is the non-dimensionalized damping matrix. Now assuming that we can write

$$[C] = \alpha[M] + \beta[K] \tag{11.20}$$

where α and β are some constants, then equation 11.19 becomes

$$m[M]\{\ddot{x}\} + c\{\alpha[M] + \beta[K]\}\{\dot{x}\} + k[K]\{x\} = \{F\}\, f/d \tag{11.21}$$

If we now repeat the process as described between equations 11.14 and 11.17b we arrive at

$$\ddot{q}_j + \{\alpha + \beta\lambda_j\}(c/m)\dot{q}_j + \lambda_j(k/m)q_j = (1/m_j)\{\psi_j\}^T\{F\}\, f/md \qquad j = 1, 2 \tag{11.22a}$$

Also, if we write

$$2\xi_j\omega_j = (\alpha + \beta\lambda_j)c/m \qquad j = 1, 2$$

and remembering that $\lambda_j = k_j/m_j = \omega_j^2 m/k$, then equation 11.22a takes the familiar form

$$\ddot{q}_j + 2\xi_j\omega_j\dot{q}_j + \omega_j^2 q_j = (\omega_j^2/k_j)\{\psi_j\}^T\{F\}\, f/kd \qquad j = 1, 2 \tag{11.22b}$$

where ξ_j is the **modal damping ratio** for the jth mode of vibration,

$$\xi_j = \frac{(c/m)(\alpha + \beta\lambda_j)}{2\omega_j} \qquad j = 1, 2 \tag{11.23}$$

Example 11.2

For the torsional system shown in Figure 11.4, $T_1 = 2\cos 30t$ and $T_2 = 3\sin 60t$. Describe the steady-state response of both rotors when the units are: polar mass moment of inertia $m = 1\,\text{kg m}^2$, torsional stiffness $k = 1000\,\text{N m/rad}$, torque $f = 1\,\text{N m}$, and displacement $d = 1\,\text{rad}$. Damping is assumed to be proportional, of the form $[C] = 3[K]$, and the unit c is $1\,\text{kg m}^2/\text{s}$.

SOLUTION

The reader should note that the system shown in Figure 11.4 is a torsional equivalence

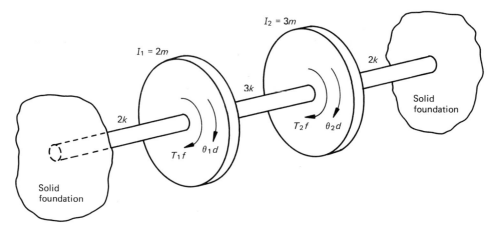

Figure 11.4

of that of Figure 11.3. Hence, as before,

$$\lambda_1 = 0.7897 \qquad \omega_1 = 28.1 \text{ rad/s} \qquad \{\psi_1\} = \{1 \quad 1.1402\}^T$$
$$\lambda_2 = 3.377 \qquad \omega_2 = 58.11 \text{ rad/s} \qquad \{\psi_2\} = \{1 \quad -0.5847\}^T$$

and $m_1 = 5.9$, $m_2 = 3.126$, $k_1 = 4.6589$ and $k_2 = 10.2176$ as before.

From equation 11.23 we have $\xi_1 = 0.04215$ and $\xi_2 = 0.087$, and from the appropriate form of equation 11.22b we have

$$\ddot{q}_1 + 2.3688\dot{q}_1 + 789.7q_1 = 0.1695(2 \cos 30t + 3.4206 \sin 60t)$$
$$\ddot{q}_2 + 10.11\dot{q}_2 + 3377q_2 = 0.3305(2 \cos 30t - 1.7541 \sin 60t)$$

Subsequently solving for the steady-state values of q_1 and q_2 gives

$$q_1 = [2.582 \cos(30t - 2.566) + 0.204 \sin(60t - 3.09)] \times 10^{-3}$$
$$q_2 = [0.265 \cos(30t - 0.122) - 0.897 \sin(60t - 1.923)] \times 10^{-3}$$

and as before θ_1 and θ_2 may be obtained from equation 11.14, with $\{\theta\}$ replacing $\{x\}$, producing

$$\theta_1 = [2.385 \cos(30t - 2.495) + 0.838 \sin(60t - 4.839)] \times 10^{-3}$$
$$\theta_2 = [3.0644 \cos(30t - 2.6) + 0.625 \sin(60t - 2.26)] \times 10^{-3}$$

(ii) *Non-proportional viscous damping.* In this case the matrix $[C]$ is not proportional to either of the matrices $[M]$ and $[K]$, nor to a linear combination of both. Accepting this to be the case, let us once again consider equation 11.19 as applied to a general n degree of freedom system. Also, let us assume that all of the components contained within the force vector $\{F\}$ are of the same harmonic frequency, Ω, but may have differing phase lags. Therefore, if we denote the components of $\{F\}$ as $F_j(j=1$ to $n)$, then we can describe them thus:

$$F_j = \hat{F}_j \cos(\Omega t + \phi_j) \qquad j = 1, n$$

which may be expressed as

$$F_j = (F_j^r + iF_j^i)\cos\Omega t$$

where i is the complex conjugate, $\sqrt{-1}$, $F_j^r = \hat{F}_j\cos\phi_j$ and $F_j^i = \hat{F}_j\sin\phi_j$.

Hence the force vector $\{F\}$ can be written as

$$\{F\} = (\{F^r\} + i\{F^i\})\cos\Omega t \tag{11.24}$$

Similarly we can describe the unknown response vector $\{x\}$ as

$$\{x\} = (\{X^r\} + i\{X^i\})\cos\Omega t \tag{11.25}$$

Therefore substituting equations 11.24 and 11.25 into equation 11.19 and equating real and imaginary terms gives

$$-\Omega^2 m[M]\{X^r\} - \Omega c[C]\{X^i\} + k[K]\{X^r\} = \{F^r\}\ f/d$$
$$-\Omega^2 m[M]\{X^i\} + \Omega c[C]\{X^r\} + k[K]\{X^i\} = \{F^i\}\ f/d \tag{11.26}$$

and expressing in matrix form gives

$$\begin{bmatrix} [K] - \Omega^2[M]m/k & \vdots & -\Omega[C]c/k \\ +\Omega[C]c/k & \vdots & [K] - \Omega^2[M]m/k \end{bmatrix} \begin{Bmatrix} \{X^r\} \\ \{X^i\} \end{Bmatrix} = \begin{Bmatrix} \{F^r\} \\ \{F^i\} \end{Bmatrix} f/kd \tag{11.27}$$

The matrix on the left-hand side of equation 11.27 is termed the **dynamic stiffness matrix** and is of dimensions $(2n \times 2n)$. For specified forcing *at a common frequency*, the vectors $\{X^r\}$ and $\{X^i\}$ can be solved for with the aid of a suitable computer subroutine, and, having done so, substituting for these vectors in equation 11.25 produces the complete response vector $\{x\}$.

In cases where the components contained in the force vector, $\{F\}$, have differing harmonic frequencies, equation 11.27 has to be solved for each value of Ω and corresponding force vectors, $\{F^r\}$ and $\{F^i\}$. Subsequently, the complete response will then be described by the sum of all the individual responses at the various frequencies, i.e. superposition.

11.3.2 Transient response of damped systems

In this case we shall consider only systems where the viscous damping is assumed to be proportional, as described by equation 11.20. In problems where the damping is non-proportional, recourse has to be made to linear multistep (LMS) methods which are beyond the scope of this text. However, an excellent review of such methods has been compiled by T. J. R. Hughes and T. Belytschko in 'A précis of developments in computational methods for transient analysis', *J. Appl. Mech., Trans. ASME*, **50**(4b), 1983, 1033–1041.

Consider therefore a general n degree of freedom system with proportional damping set into transient motion by an initial displacement vector, $\{x\}_{int}$, and/or an initial velocity vector, $\{\dot{x}\}_{int}$. Now prior to applying equation 11.22b with the harmonic force vector $\{F\} = \{0\}$ for transient analysis, it is necessary to transform the initial displacement and velocity vectors into the corresponding modal coordinate vectors, $\{q\}_{int}$ and $\{\dot{q}\}_{int}$ respectively. From equation 11.14 we have

$$\{x\}_{int} = [V]\{q\}_{int} \tag{11.28}$$

and

$$\{\dot{x}\}_{int} = [V]\{\dot{q}\}_{int}$$

If we now pre-multiply both sides of equation 11.28 by $[V]^T[M]$ we have

$$[V]^T[M]\{x\}_{int} = [M]\{q\}_{int}$$

and   (11.29)

$$[V]^T[M]\{\dot{x}\}_{int} = [M]\{\dot{q}\}_{int}$$

Since the modal mass matrix $[M]$ is diagonal, the inversion of such is a diagonal matrix whose diagonal terms are the reciprocal of the modal mass values, $1/m_j$, $j = 1, n$. Hence

$$\{q\}_{int} = [M]^{-1}[V]^T[M]\{x\}_{int}$$

and   (11.30)

$$\{\dot{q}\}_{int} = [M]^{-1}[V]^T[M]\{\dot{x}\}_{int}$$

Example 11.3

For the system described in Example 11.2, in the absence of the harmonic forcing, describe the transient response of the system when the $3m$ rotor is given an initial static displacement of 0.035 rad (the $2m$ rotor held rigidly) and then released.

SOLUTION

$$[M]^{-1}[V]^T[M] = \begin{bmatrix} \frac{1}{5.9} & 0 \\ 0 & \frac{1}{3.1256} \end{bmatrix} \begin{bmatrix} 1 & 1.1402 \\ 1 & -0.5847 \end{bmatrix} \begin{bmatrix} 2 & 0 \\ 0 & 3 \end{bmatrix}$$

$$= \begin{bmatrix} 0.339 & 0.5798 \\ 0.64 & -0.5612 \end{bmatrix}$$

and

$$\{x\}_{int} = \{0 \quad 0.035\}^T \qquad \{\dot{x}\}_{int} = \{0 \quad 0\}^T$$

Hence, from equation 11.30,

$$\{q\}_{int} = \begin{Bmatrix} q_1 \\ q_2 \end{Bmatrix}_{int} = \begin{Bmatrix} 0.0203 \\ -0.0196 \end{Bmatrix} \quad \text{and} \quad \{\dot{q}\}_{int} = \begin{Bmatrix} \dot{q}_1 \\ \dot{q}_2 \end{Bmatrix}_{int} = \begin{Bmatrix} 0 \\ 0 \end{Bmatrix}$$

Now from equation 11.22b, with the torque vector $\{T\}$ (replacing $\{F\}$ in this case) $= \{0\}$, we have

$$\ddot{q}_j + 2\xi_j\omega_j\dot{q}_j + \omega_j^2 q_j = 0 \qquad j = 1, 2$$

where, as before $\xi_1 = 0.04215$, $\xi_2 = 0.087$, $\omega_1 = 28.1$ rad/s and $\omega_2 = 58.1$ rad/s. Hence

$$\ddot{q}_1 + 2.3688\dot{q}_1 + 789.7q_1 = 0$$
$$\ddot{q}_2 + 10.11\dot{q}_2 + 3377q_2 = 0$$

The general solutions to which (see Section 9.4.1) are

$$q_1 = A_1 \exp(-1.1844t)\cos(28.075t + \varepsilon_1)$$
$$q_2 = A_2 \exp(-5.055t)\cos(57.92t + \varepsilon_2)$$

where A_1, A_2, ε_1 and ε_2 are constants determined by the initial conditions, $\{q\}_{int}$

and $\{\dot{\mathbf{q}}\}_{int}$, giving

$$\mathbf{q}_1 = 0.02032 \exp(-1.1844t)\cos(28.075t - 0.04216)$$
$$\mathbf{q}_2 = 0.0197 \exp(-5.055t)\cos(57.92t - 3.2287)$$

Then equation 11.14, with $\{\theta\}$ replacing $\{x\}$, gives

$$\theta_1 = 0.02032 \exp(-1.1844t)\cos(28.075t - 0.04216)$$
$$+ 0.0197 \exp(-5.055t)\cos(57.92t - 3.2287)$$

$$\theta_2 = 0.02317 \exp(-1.1844t)\cos(28.075t - 0.04216)$$
$$- 0.0115 \exp(-5.055t)\cos(57.92t - 3.2287)$$

11.4   Experimental determination of modal parameters

So far, in our study of modal analysis, we have assumed throughout that values of mass and stiffness have been known, and that on the basis of such values the transient response and steady-state response to harmonic forcing could be predicted. In practice, however, the *true* values of mass and stiffness associated with a system or structure are often very much different from the values that may have been assumed, resulting in the true response of the system differing from that predicted on the basis of the assumed values of mass and stiffness. Furthermore, in the absence of conducting experimental testing on the actual system, the degree of damping present is almost impossible to predict. It is for these reasons that the experimental aspect of modal analysis can become extremely useful in determining close estimates of the true values of the modal mass, stiffness and damping ratios associated with the system. Although a rigorous treatise of the components and procedures associated with experimental modal analysis are beyond the scope of this text, their importance cannot be over-emphasized and the reader, prior to embarking on such an experimental investigation, is encouraged to consult the comprehensive text on this subject by D.J. Ewins (*Modal Testing: Theory and Practice*, Research Studies Press, 1984). However, for the present we shall assume that if a system is subjected to controlled, variable frequency, harmonic forcing, then:

(1)   At a resonance condition, i.e. where the excitation frequency is equal to one of the undamped natural frequencies of the system, the vibratory response at the point on the system where the harmonic forcing is applied is 90° out of phase (lagging) with the harmonic forcing, see Section 9.5.1.

(2)   At such a resonance condition, for a particular normal mode of vibration, the amplitudes of the relative vibratory displacement at points on the system are representative of the eigenvector associated with the same normal mode. This can be shown to be true.

(3)   At a resonance condition of one particular normal mode of vibration, the vibratory contribution from other normal modes is negligible. This assumption is feasible for systems whose natural frequencies are well separated numerically. However, for systems whose natural frequencies tend to be close together, compensatory techniques can be applied as demonstrated by C.C. Kennedy and C.D.P. Pancu ('Use of vectors in vibration measurement and analysis', *J. Aeronautical Sci.* **14**(11), 1947, 603–625) and C.V. Stahle Jr ('Phase separation technique for ground vibration testing', *Aerospace Eng.*, July 1962).

(4) Damping is basically viscous and proportional. Also, from the analysis of Section 9.5.1, for small amounts of damping the value of the damping ratio, ξ, for a particular normal mode of vibration can be obtained from the frequency response plot, around the mode being considered, of a point on the system (other than a node point), i.e. see Figure 11.5, such that

$$\xi = \frac{\Omega_a - \Omega_b}{\Omega_a + \Omega_b}$$

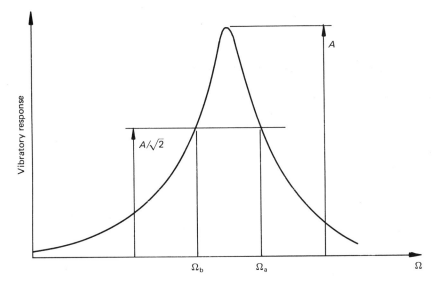

Figure 11.5

(5) The stiffness of the system is basically linear.

On the basis of statements 1 to 5 above let us now proceed to demonstrate, by means of a numerical example, what can be obtained from the results of such experimental testing carried out on a system.

Example 11.4

Experimental modal testing was carried out on the two degree of freedom torsional system shown in Figure 11.6. The testing comprised applying a harmonic torque at rotor 1 of amplitude 1 N m at each of the two undamped natural frequencies of the system, ω_1 and ω_2. In each case the vibratory amplitude of both rotors, $\hat{\theta}_1$ and $\hat{\theta}_2$, were measured and the appropriate modal damping ratios, ξ_1 and ξ_2, were determined from the procedure described in statement 4 above. The results obtained from this testing are given in Table 11.1, where the minus sign indicates that θ_2 was found to be 180° out of phase with θ_1.

Table 11.1

Mode j	ω_j (rad/s)	$\hat{\theta}_1$ (rad)	$\hat{\theta}_2$ (rad)	ξ_j
1	56	0.035	0.02	0.01
2	184	0.01	−0.015	0.108

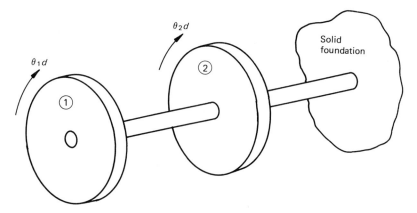

Figure 11.6

On the basis of these results, estimate and describe the vibratory response of the system when operating under normal condition such that a harmonic torque, having an amplitude and frequency of 1.5 N m and 50 Hz, respectively, is applied to rotor 2.

SOLUTION

From the vibratory amplitudes normalized to $\hat{\theta}_1$ and from statement 2 above, the modal matrix [V] is

$$[\mathbf{V}] = \begin{bmatrix} 1 & 1 \\ 0.5714 & -1.5 \end{bmatrix}$$

and the torque vector $\{T\}$ which replaces the force vector $\{F\}$ in equation 11.22b is

$$\{T\} = \{1 \quad 0\}^{\mathrm{T}} \cos \Omega t$$

where Ω is the applied frequency.

For the first natural frequency ($j = 1$) and $\Omega = \omega_1$, we have from the solution of equation 11.22b

$$\mathbf{q}_1 = \frac{50}{\mathbf{k}_1} \cos(\omega_1 t - \pi/2) \, \text{N m}/k. \, \text{rad}$$

Since the two natural frequencies are well separated, from statement 3 above, $\mathbf{q}_2$

can be assumed to be zero, although the reader may wish to verify that in actual fact

$$\mathbf{q}_2 = \frac{1.099}{\mathbf{k}_2} \cos(\omega_1 t - 0.0723) \, \text{N m}/k. \text{ rad}$$

Therefore assuming $\mathbf{q}_2$ to be zero we have from $\{\theta\} = [\mathbf{V}]\{\mathbf{q}\}$, and remembering that $\hat{\theta}_1 = 0.035$,

$$\mathbf{k}_1 = \frac{50}{0.035} = 1428.57$$

Now, assuming k and m to be N m/rad and kg m^2 respectively,

$$\mathbf{m}_1 = \frac{\mathbf{k}_1}{\omega_1^2} = 0.4555$$

Similarly, if we now repeat the same exercise for the second normal mode ($j = 2$, $\Omega = \omega_2$), we deduce that $\mathbf{k}_2 = 462.96$ and $\mathbf{m}_2 = 0.01367$.

Under the normal operating conditions described, the torque vector $\{T\}$ becomes

$$\{T\} = \{0 \quad 1.5\}^\text{T} \cos 100 \pi t$$

Therefore solving equation 11.22b with $\{T\}$ replacing $\{F\}$ gives

$$\mathbf{q}_1 = 0.0197 \cos(100 \pi t - 3.138) \times 10^{-3}$$

and

$$\mathbf{q}_2 = -2.492 \cos(100 \pi t - 2.9513) \times 10^{-3}$$

Hence, from $\{\theta\} = [\mathbf{V}]\{\mathbf{q}\}$ and simplifying, we obtain

$$\theta_1 = 2.473 \cos(100 \pi t + 0.1918) \times 10^{-3}$$

and

$$\theta_2 = 3.749 \cos(100 \pi t - 2.9513) \times 10^{-3}$$

Problems

11.1 For the torsional system shown in Figure 11.7, show that the eigenvalues and corresponding eigenvectors are

Eigenvalue $= m\omega^2/k$	Eigenvector $= \{\hat{\theta}_1 \quad \hat{\theta}_2\}^\text{T}$
0.1396	$\{1 \quad\quad 0.5812\}^\text{T}$
1.1937	$\{1 \quad -2.5811\}^\text{T}$

where ω is the undamped natural frequency and k and m are 1 N m/rad and 1 kg m^2 respectively. Hence calculate the steady-state response of both rotors when the $3m$ rotor is subjected to a harmonic torque of the form $T_0(t) = 1 \cos 5t$ N m. Assume that damping is proportional and of the form $[C] = 0.05 [K]$, where $[C]$ and $[K]$ are the non-dimensionalized damping and stiffness matrices respectively, and assume that the unit of damping, c, is 1 kg m^2/s.

Answer: $\theta_1 = 0.0137 \underline{/-180°}$; $\theta_2 = -0.00018 \underline{/-180°}$.

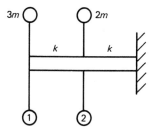

Figure 11.7

11.2   For the linear system shown in Figure 11.8, show that the eigenvalues and corresponding eigenvectors are

Eigenvalue $m\omega^2/k$	Eigenvector $= \{X_1 \quad X_2\}^T$
1.396	$\{1 \qquad 1.104\}^T$
3.104	$\{1 \quad -0.604\}^T$

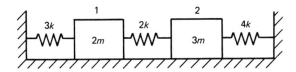

Figure 11.8

where ω is the undamped natural frequency and k and m are 1 N/m and 0.1 kg respectively. Hence calculate the steady-state response of both bodies when the $3m$ body is subjected to a harmonic force of the form, $F(t) = 0.01 \sin 5t$ N. Assume that damping is proportional and of the form $[C] = 0.2\,[K]$, where $[C]$ and $[K]$ are the non-dimensionalized damping and stiffness matrices respectively, and assume that the unit of damping, c, is 1 kg/s.

Answer: $x_1 = \{1.09 \underline{/-128°} - 0.617 \underline{/-79°}\} \times 10^{-3}$
$\qquad x_2 = \{1.2 \underline{/-128°} - 0.373 \underline{/-79°}\} \times 10^{-3}$

11.3   Resonance testing was carried out on the two degree of freedom torsional system shown in Figure 11.9. The test comprised applying a harmonic torque of constant amplitude of 1 N m at rotor 2 at each of the two undamped natural frequencies of the system, ω_1 and ω_2. In each case the peak vibratory amplitude

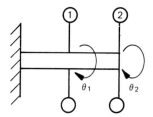

Figure 11.9

at both rotors was measured together with the appropriate damping ratios, ξ_1 and ξ_2. The results of this test are:

Mode n	$\omega_n(rad/s)$	$\hat{\theta}_1(rad)$	$\hat{\theta}_2(rad)$	ξ_n
1	56	0.02	0.035	0.01
2	84	−0.015	0.01	0.015

Determine the modal stiffness values, $k_j (j = 1, 2)$, and modal inertia values m_j. State clearly any assumptions you make.

Answer: $k_1 = 4375\,N\,m/rad$; $m_1 = 1.395\,kg\,m^2$, $k_2 = 1482\,N\,m/rad$; $m_2 = 0.21\,kg\,m^2$.

12

Free vibration of continuous systems

12.1 Introduction

In the previous chapters dealing with vibration, the systems of bodies performing vibratory motion were considered to be discrete or lumped systems, and modelled accordingly, i.e. the constituent parts of the system were treated as idealized point masses or rigid bodies connected by essentially massless elastic units. Such systems possess a finite number of degrees of freedom and have, therefore, a finite number of natural frequencies and associated vibratory modes. Many practical problems lend themselves to such forms of analysis but there are disadvantages in that mass and elasticity are not always easily separated in real systems. Elements such as shafts, beams, cables, plates and shells are examples that fall into this category, creating difficulties in terms of rigorous solution using discretization. An alternative form of modelling which can be used is that of distributed mass and elasticity, where these parameters are considered to be continuously distributed throughout the system. In this chapter we shall confine our study to the free response of some of the more commonly encountered *one-dimensional* distributed systems, i.e. where the vibration is a function of only one spatial coordinate. Furthermore, we shall only consider systems where

(1) the constituent material is homogeneous, isotropic and obeys Hooke's law,
(2) the displacements are sufficiently small to ensure a linear-elastic response, and
(3) damping is negligible.

12.2 Transverse vibration of a string or cable

Perhaps the simplest of all continuous media is that of a string or cable drawn taut by the action of a constant tension. Initial transverse displacement and subsequent release of the string will induce vibratory motion at one or more of the many natural frequencies of the system.

For ease of solution the string is assumed to have negligible flexural rigidity. Consider such a string, shown in Figure 12.1a, having uniform mass distribution, m per length, subjected to a constant tension, s, which is unaffected by small lateral displacements of the string.

If gravitational effects are neglected, the system of forces acting on an elemental

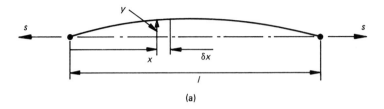

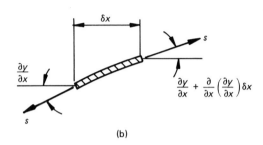

Figure 12.1

length, δx, of the string will be as shown in Figure 12.1b, and the net effective vertical force δF acting on this element is

$$\delta F = s\left(\frac{\partial y}{\partial x} + \frac{\partial^2 y}{\partial x^2}\cdot\delta x\right) - s\frac{\partial y}{\partial x} \tag{12.1}$$

or

$$\delta F = s\frac{\partial^2 y}{\partial x^2}\cdot\delta x \tag{12.2}$$

Applying Newton's Second Law only to the *rectilinear vertical motion* of the element gives

$$\delta F = \frac{\partial}{\partial t}\left(m\,\delta x\cdot\frac{\partial y}{\partial t}\right) \tag{12.3}$$

Substituting for δF in equation 12.2 gives

$$\frac{\partial^2 y}{\partial x^2} = \frac{1}{c^2}\frac{\partial^2 y}{\partial t^2} \tag{12.4}$$

where $c = \sqrt{(s/m)}$ represents the longitudinal velocity of transverse waves on the string. Equation 12.4 is referred to as the one-dimensional wave equation of which there are many solutions; however, the one most applicable to the free vibration of continuous systems is the harmonic solution, where the system is considered to be performing a normal mode of vibration at one of its natural frequencies, i.e.

$$y = y(x, t) = Y_n(x)(A_n \cos \omega_n t + B_n \sin \omega_n t) \tag{12.5}$$

where $Y_n(x)$ is the normal function representing the nth vibratory mode, A_n and B_n are constants determined from the *initial conditions* governing a specific problem, and ω_n is the nth natural frequency. Substituting for y in equation 12.4 gives

$$\frac{\mathrm{d}^2 Y_n(x)}{\mathrm{d}x^2} + \frac{\omega_n^2}{c^2} Y_n(x) = 0 \tag{12.6}$$

the solution to which is

$$Y_n(x) = C_n \cos(\omega_n x/c) + D_n \sin(\omega_n x/c) \tag{12.7}$$

where C_n and D_n are constants which may be determined from the *boundary conditions* that exist at the ends of the string.

Example 12.1

Determine the natural frequencies and associated modes of lateral vibration that can exist on a string of uniform mass, rigidly held at both ends and subject to a constant tension.

SOLUTION

The boundary conditions for this problem are such that

$$y(0, t) = y(l, t) = 0$$

i.e.

$$Y_n(0) = Y_n(l) = 0 \qquad \text{for all values of } n$$

Substituting in equation 12.7 gives

$$0 = C_n \tag{12.8}$$

and

$$0 = D_n \sin(\omega_n l/c) \tag{12.9}$$

Neglecting the trivial solution for equation 12.9, i.e. $D_n = 0$, the equation is satisfied by the relationship

$$\frac{\omega_n l}{c} = n\pi \tag{12.10}$$

where $n = 1, 2, 3 \ldots \infty$, and ω_n becomes

$$\omega_n = \frac{n\pi}{l} \sqrt{\frac{s}{m}} \tag{12.11}$$

Substituting for ω_n in equation 12.7 gives the deflected form of the string for the nth mode, i.e.

$$Y_n(x) = D_n \sin(n\pi x/l) \tag{12.12}$$

Figure 12.2 shows the first three modes that can exist on the taut string. If all the modes were excited simultaneously, the complete response of the string would be the sum of all the individual responses, i.e.

$$y(x, t) = \sum_{n=1}^{\infty} D_n \sin\left(\frac{n\pi x}{l}\right)(A_n \cos \omega_n t + B_n \sin \omega_n t) \tag{12.13}$$

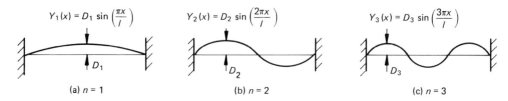

$$Y_1(x) = D_1 \sin\left(\frac{\pi x}{l}\right) \qquad Y_2(x) = D_2 \sin\left(\frac{2\pi x}{l}\right) \qquad Y_3(x) = D_3 \sin\left(\frac{3\pi x}{l}\right)$$

(a) $n = 1$ (b) $n = 2$ (c) $n = 3$

Figure 12.2

PRACTICAL NOTE

In steam turbines, for the purpose of reinforcement, the blades are sometimes wired together in groups as shown in Figure 12.3. The flow of steam over the wires may often induce lateral vibration of the wires (aeolian tones) and if the frequency of this vibration (or any of its harmonics) is close to any of the natural frequencies of the turbine blades, large vibratory amplitudes of the latter may ensue, culminating in failure of the blades if this resonance condition is left unabated.

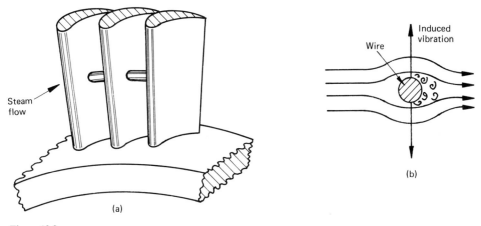

(a)

(b)

Figure 12.3

12.3 Longitudinal vibration of a prismatic bar

Another form of vibratory motion associated with distributed mass systems is the compression/extension wave motion which can take place along the length of a prismatic elastic beam or bar. In this case it is assumed that beam cross-sections remain plane and that the particles in these plane sections execute motion only in the longitudinal or axial direction, i.e. there is no lateral shear force action. Consider the bar. shown in Figure 12.4a, having length l and uniform constant cross-sectional area A.

At some position x along the bar, the axial force system acting on a longitudinal element of length δx will be as shown in Figure 12.4b, where σ is the longitudinal

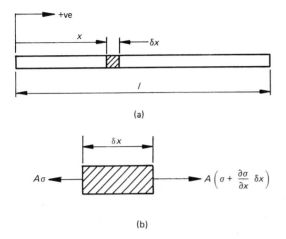

Figure 12.4

stress at section x. Let u be the longitudinal displacement of any cross-section located at a distance x from the origin, ε the strain $(\partial u/\partial x)$, E the Young's modulus of elasticity, and ρ the mass density of bar material.

Now the net force (either tensile or compressive) δf acting on element δx will be

$$\delta f = A\left(\sigma + \frac{\partial \sigma}{\partial x}\delta x\right) - A\sigma$$

or

$$\delta f = A\frac{\partial \sigma}{\partial x}\delta x \tag{12.14}$$

However

$$\sigma = E\varepsilon \qquad \text{and} \qquad \varepsilon = \frac{\partial u}{\partial x}$$

Therefore

$$\frac{\partial \sigma}{\partial x} = E\frac{\partial^2 u}{\partial x^2}$$

Substituting in equation 12.14 gives the net force

$$\delta f = AE\frac{\partial^2 u}{\partial x^2}\delta x$$

and applying Newton's Second Law to the element gives

$$\delta f = AE\frac{\partial^2 u}{\partial x^2}\delta x = \rho A\delta x\frac{\partial^2 u}{\partial t^2}$$

which simplifies to

$$\frac{\partial^2 u}{\partial x^2} = \frac{\rho}{E}\frac{\partial^2 u}{\partial t^2} \tag{12.15}$$

Since the velocity of wave propagation in a solid elastic material is $c = \sqrt{(E/\rho)}$ (i.e. the speed of sound in that material), the equation of motion governing the longitudinal wave motion in the bar becomes

$$\frac{\partial^2 u}{\partial x^2} = \frac{1}{c^2}\frac{\partial^2 u}{\partial t^2} \tag{12.16}$$

which is of identical form to that of equation 12.4; that is to say it represents the one-dimensional wave equation, the solution of which is given by equation 12.5 with the lateral displacement $y(x, t)$ being replaced by the longitudinal displacement $u(x, t)$. Thus for the nth mode

$$u(x, t) = U_n(x)(A_n \cos \omega_n t + B_n \sin \omega_n t) \tag{12.17}$$

The corresponding form of the nominal function $U_n(x)$, depicting the mode of longitudinal vibratory displacement, will be that given by equation 12.7. Once again the constants A_n and B_n will be dependent on the manner in which vibration is initiated, whilst C_n and D_n are controlled by the governing boundary conditions, usually at the end of the bar.

Example 12.2

(a) Determine the natural frequencies and associated modes of longitudinal vibration that can occur in a uniform bar which is fixed at one end and free at the other.

(b) If the free end of the bar is subjected to a static load and then suddenly released, determine the subsequent response of the bar.

SOLUTION

(a) As has been shown previously, the solution to the wave equation governing such problems is

$$u(x, t) = \left[C_n \cos\left(\frac{\omega_n x}{c}\right) + D_n \sin\left(\frac{\omega_n x}{c}\right) \right][A_n \cos \omega_n t + B_n \sin \omega_n t] \tag{12.18}$$

The geometric boundary conditions that apply to this problem are that at the fixed end $(x = 0)$ the longitudinal displacement is zero and at the free end $(x = l)$ there can be no externally applied force during the free vibration, i.e. the longitudinal strain is zero; thus we have the relationships

$$u(0, t) = 0 \tag{12.19}$$

$$\varepsilon(l, t) = \frac{\partial u}{\partial x}(l, t) = 0 \tag{12.20}$$

Substitution of $u(x, t)$ from equation 12.18 in equation 12.19 gives $C_n = 0$. Differentiating equation 12.18 with respect to x and substituting in equation 12.20 gives the relationship

$$\frac{\partial u}{\partial x}(l,t) = 0 = D_n \frac{\omega_n}{c} \cos\left(\frac{\omega_n l}{c}\right)(A_n \cos \omega_n t + B_n \sin \omega_n t) \qquad (12.21)$$

The non-trivial solution of equation 12.21 corresponds to

$$\cos(\omega_n l/c) = 0$$

which is true for $\omega_n l/c = \frac{\pi}{2}(2n-1)$ where $n = 1, 2, 3$, etc. Thus the corresponding natural frequencies are

$$\omega_n = \frac{(2n-1)\pi c}{2l}$$

or

$$\omega_n = \frac{(2n-1)\pi}{2l}\sqrt{\frac{E}{\rho}} \qquad (12.22)$$

The longitudinal displacement associated with the nth normal mode is therefore

$$u(x,t) = D_n \sin\left[\frac{(2n-1)\pi x}{2l}\right] \cdot [A_n \cos \omega_n t + B_n \sin \omega_n t] \qquad (12.23)$$

and the complete response of the bar if all possible modes are excited simultaneously will be

$$u(x,t) = \sum_{n=1}^{\infty} D_n \sin\left[\frac{(2n-1)\pi x}{2l}\right] \cdot [A_n \cos \omega_n t + B_n \sin \omega_n t] \qquad (12.24)$$

(b) We can use the results of the above to solve the second part of the problem. If the bar is subjected to an initial static force, F, applied at the free end (i.e. at $x = l$), the corresponding initial strain will be

$$\varepsilon_0 = F/AE$$

Also, the longitudinal displacement at any point along the bar will be linearly proportional to the distance from the fixed end, i.e. $u(x,0) = \varepsilon_0 x$. Additionally, prior to the withdrawal of the force, all points on the bar will be stationary, i.e.

$$\frac{\partial u}{\partial t}(x,0) = 0$$

After differentiating equation 12.24, this latter condition will be seen to be satisfied by making $B_n = 0$. Substituting in equation 12.23 for $B_n = 0$ and $t = 0$ results in the relationship

$$A_n D_n \sin\left[\frac{(2n-1)\pi x}{2l}\right] = \varepsilon_0 x$$

or

$$A_n' \sin\left[\frac{(2n-1)\pi x}{2l}\right] = \varepsilon_0 x \qquad (12.25)$$

where $A'_n = A_n D_n$. From Fourier analysis, utilizing the relevant coefficients, we have

$$A'_n = \frac{2\varepsilon_0}{l} \int_0^l x \sin\left[\frac{(2n-1)\pi x}{2l}\right] dx$$

which after integration by parts results in

$$A'_n = \frac{8\varepsilon_0 l}{(2n-1)^2 \pi^2} \sin\left[\frac{(2n-1)\pi}{2}\right]$$

Since for this particular problem the integer $(2n-1)$ can only be odd, it follows that

$$\sin\frac{(2n-1)\pi}{2} = \pm 1 \qquad \text{or} \qquad \sin\frac{(2n-1)\pi}{2} = (-1)^{n-1}$$

giving

$$A'_n = \frac{8\varepsilon_0 l}{(2n-1)^2 \pi^2}(-1)^{n-1} \tag{12.26}$$

The complete solution is therefore

$$u(x,t) = \frac{8\varepsilon_0 l}{\pi^2} \sum_{n=1}^{\infty} \frac{(-1)^{n-1}}{(2n-1)^2} \sin\left[\frac{(2n-1)\pi x}{2l}\right] \cos \omega_n t \tag{12.27}$$

The amplitudes of longitudinal vibration corresponding to the free end of the bar, for the first three modes, will be in the ratio 1, 1/9, 1/25, etc.; i.e. diminishing as the inverse square of $(2n-1)$ where n is the mode number, i.e. 1, 2, 3, etc.

12.4 Torsional vibration of a uniform circular bar

In Chapter 10 we examined the torsional vibration characteristics of rotor systems, where the mass of the elastic shaft connecting the rotors was neglected. Such shafts, however, in themselves represent uniformly distributed mass/elastic systems possessing an infinite number of natural modes of free torsional oscillation and, although the frequencies associated with these motions are generally much higher than those of the discrete rotor systems, they can on occasions cause problems.

Consider the straight shaft, of length l, having a circular cross-section of area A, shown in Figure 12.5a. Let θ be the rotation of any cross-section located at distance x from one end of the shaft as it performs torsional vibratory motion about its central axis. During this motion the torque applied to an element δx will be as shown in Figure 12.5b. Thus the net elemental torque δT acting on the element is

$$\delta T = T + \frac{\partial T}{\partial x}\delta x - T = \frac{\partial T}{\partial x}\delta x \tag{12.28}$$

Applying Newton's Second Law (in its angular form) gives

$$\frac{\partial T}{\partial x}\delta x = I\frac{\partial^2 \theta}{\partial t^2} \tag{12.29}$$

where I is the polar mass moment of inertia of the element $(\rho J_0 \delta x)$, ρ is the mass density of the shaft material, and J_0 is the polar second moment of area of shaft section.

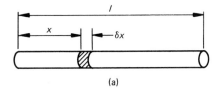

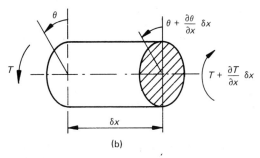

Figure 12.5

Substituting in equation 12.29 gives the relationship

$$\frac{\partial T}{\partial x} = \rho J_0 \frac{\partial^2 \theta}{\partial t^2} \qquad (12.30)$$

Now from elementary theory for the twisting of a uniform shaft we have

$$T/J_0 = \tau/r = G\phi/l$$

where τ is the shear stress at radius r, G is the modulus of rigidity of the shaft material, and ϕ is the total twist of shaft over length l.

The above relationship may be rewritten as

$$T/GJ_0 = \phi/l = \text{angular displacement per length of shaft}$$

For the element of length δx the net twist will be

$$\theta + \frac{\partial \theta}{\partial x} \delta x - \theta = \frac{\partial \theta}{\partial x} \delta x$$

and the twist per unit length $= \partial\theta/\partial x$.

Thus the basic equation governing the twisting of the element becomes

$$T = GJ_0 \frac{\partial \theta}{\partial x} \qquad (12.31)$$

Differentiating equation 12.31 with respect to x gives

$$\frac{\partial T}{\partial x} = GJ_0 \frac{\partial^2 \theta}{\partial x^2} \qquad (12.32)$$

and substituting in equation 12.30 results in

$$\frac{\partial^2 \theta}{\partial x^2} = \frac{\rho}{G} \cdot \frac{\partial^2 \theta}{\partial t^2} \qquad (12.33)$$

which may be expressed in the form

$$\frac{\partial^2 \theta}{\partial x^2} = \frac{1}{c^2} \cdot \frac{\partial^2 \theta}{\partial t^2} \tag{12.34}$$

Equation (12.34) is once again the one-dimensional wave equation; in this case $c = \sqrt{(G/\rho)}$ represents the velocity of propagation of shear stress waves along the bar. The solution to equation 12.34 is, of course, given by equation 12.5 when the variable $y(x,t)$ is replaced by $\theta(x,t)$.

Thus all three forms of oscillatory motion, i.e. lateral, longitudinal and angular, associated with the distributed mass/elastic systems analysed, are essentially similar in terms of solution.

12.5   Transverse vibration of a prismatic beam

In Section 12.2 we analysed the free transverse vibration of a string or wire where the restoring effort returning the string to its equilibrium position, after vibratory deflection, was produced purely by the tension; in other words the flexural stiffness or rigidity of the string was considered to be negligible. However, in a distributed mass/elastic system such as a beam, the flexural stiffness of the beam is of prime importance in the analysis of transverse vibration.

Consider the prismatic beam of length l, uniform cross-sectional area A, and second moment of area J about the neutral axis of bending as shown in Figure 12.6a.

The lateral deflection of the beam at the section distance x from the left-hand end is $y = y(x,t)$, and the system of forces and moments applied to an element of length δx, located at x, will be as shown in Figure 12.6b. As the beam flexes during transverse vibration, this element and all similar elements will deflect in the OY direction and also rotate slightly about the OZ axis. The contribution this rotation makes to the vibratory energy of the beam will be examined in Section 12.5.1, but for the moment let us neglect this rotational effect and concentrate on the lateral movement of the element.

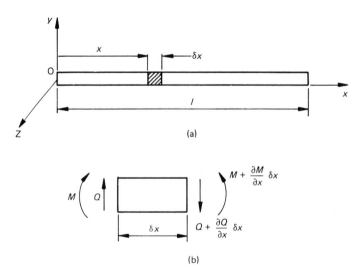

Figure 12.6

Summing the moments about the left-hand end of the element gives

$$M - \left(M + \frac{\partial M}{\partial x}\delta x\right) + \left(Q + \frac{\partial Q}{\partial x}\delta x\right)\delta x + \rho g A\,\delta x\cdot\frac{\delta x}{2} = 0 \qquad (12.35)$$

where ρ is the mass density of the beam material and g is the gravitational acceleration. Neglecting the product of small quantities, equation 12.35 reduces to

$$Q = \frac{\partial M}{\partial x} \qquad (12.36)$$

The net vertical force on the element will be

$$Q - \left(Q + \frac{\partial Q}{\partial x}\delta x\right) = -\frac{\partial Q}{\partial x}\delta x$$

and applying Newton's Second Law to the element for vertical rectilinear motion only (i.e. neglecting rotation of the element) we have

$$-\frac{\partial Q}{\partial x}\delta x = \frac{\partial}{\partial t}\left(\rho A\,\delta x\frac{\partial y}{\partial t}\right)$$

or

$$-\frac{\partial Q}{\partial x} = \rho A\frac{\partial^2 y}{\partial t^2} \qquad (12.37)$$

Differentiating equation 12.36 with respect to x and substituting in equation 12.37 gives

$$-\frac{\partial^2 M}{\partial x^2} = \rho A\frac{\partial^2 y}{\partial t^2} \qquad (12.38)$$

In addition, from elementary bending theory for a thin uniform beam, we have the relationship

$$EJ\frac{\partial^2 y}{\partial x^2} = M \qquad (12.39)$$

where E is the modulus of elasticity of the beam material, and the product EJ represents the flexural rigidity of the beam. Differentiating equation 12.39 with respect to x twice and substituting in equation 12.38 gives

$$-EJ\frac{\partial^4 y}{\partial x^4} = \rho A\frac{\partial^2 y}{\partial t^2} \qquad (12.40)$$

This is the fourth-order partial differential equation which may be solved in a similar manner to that of the one-dimensional wave equation, i.e. by separation of variables. When the beam vibrates transversely in one of its natural modes, the deflection at any point along the beam may be assumed to be equal to the product of two functions

$$y = y(x, t) = Y_n(x)\,\phi(t) \qquad (12.41)$$

where $Y_n(x)$ is the spatial or normal function depicting the modal configuration of the deflected beam and $\phi(t)$ the harmonic variation of the displacement. In a similar manner, as before, $y(x, t)$ may be expressed as

$$y(x, t) = Y_n(x)(A_n \cos \omega_n t + B_n \sin \omega_n t) \qquad (12.42)$$

where A_n and B_n are constants dependent on the conditions that initiate the vibration.

Differentiating equation 12.42 with respect to x and t and substituting in equation 12.40 produces the relationship

$$EJ\frac{d^4 Y_n(x)}{dx^4} - \rho A\omega_n^2 Y_n(x) = 0 \tag{12.43}$$

or expressed in simpler form as

$$\frac{d^4 Y_n(x)}{dx^4} - k_n^4 Y_n(x) = 0 \tag{12.44}$$

where

$$k_n = \left(\frac{\rho A\omega_n^2}{EJ}\right)^{1/4}$$

Equation 12.44 may be satisfied by assuming a solution of the form $Y_n(x) = C_n \exp(\alpha x)$, which after differentiation and substitution in equation 12.44 gives

$$(\alpha^4 - k_n^4)C\exp(\alpha x) = 0 \tag{12.45}$$

from which the non-trivial solution is

$$\alpha^4 = k_n^4$$

or

$$\alpha^2 = \pm k_n^2 \tag{12.46}$$

Consequently we have four values of α that satisfy equation 12.46, namely $\alpha_1 = k_n$, $\alpha_2 = -k_n$, $\alpha_3 = ik_n$ and $\alpha_4 = -ik_n$, where i is the complex conjugate, $\sqrt{-1}$. All four values of α combine to give the complete solution for the normal function $Y_n(x)$ as

$$Y_n(x) = C_n \exp(k_n x) + D_n \exp(-k_n x) + E_n \exp(ik_n x) + F_n \exp(-ik_n x) \tag{12.47}$$

where $C_n \ldots F_n$ are constants that may be determined (or expressed in terms of some single arbitrary constant) from the boundary conditions which exist, usually at each end of the beam, for any given problem. Substituting the relationships

$$\exp(\pm ik_n x) = \cos k_n x \pm \sin k_n x$$

and

$$e(\pm k_n x) = \cosh k_n x \pm \sinh k_n x$$

into equation 12.47 gives

$$Y_n(x) = C_n \sin k_n x + D_n \cos k_n x + E_n \sinh k_n x + F_n \cosh k_n x \tag{12.48}$$

where $C_n \ldots F_n$ are modified versions of the constants $C_n \ldots F_n$ in equation 12.47. Examples of the boundary conditions most often encountered in practical beam vibration problems are as follows:

(1) *Free end*   Bending moment M (about OZ) and shear force Q (in the OY direction) are zero at all times, i.e.

$$M = EJ\frac{\partial^2 y}{\partial x^2} = 0 \qquad \text{at all times} \tag{12.49}$$

and

$$Q = \frac{\partial M}{\partial x} = EJ\frac{\partial^3 y}{\partial x^3} = 0 \qquad \text{at all times} \tag{12.50}$$

Equation 12.49 is satisfied only by the relationship $d^2 Y_n(x)/dx^2 = 0$, and equation 12.50 by $d^3 Y_n(x)/dx^3 = 0$.

(2) *Simply supported (pinned) end*   Deflection and bending moment are zero at all times, i.e.

$$Y_n(x) = 0$$

and $M = 0$, i.e.

$$\frac{d^2 Y_n(x)}{dx^2} = 0$$

(3) *Fixed (clamped) end*   Deflection and slope are zero at all times, i.e.

$$Y_n(x) = 0 \qquad \text{and} \qquad \frac{d Y_n(x)}{dx} = 0$$

For any given problem there will always be four boundary conditions which will enable the constants C_n to F_n to be determined. However the beam, being a continuously distributed mass/elastic system, will possess an infinite number of vibratory modes, and the complete response of the beam will be given by the expression

$$y(x, t) = \sum_{n=1}^{\infty} Y_n(x)[A_n \cos \omega_n t + B_n \sin \omega_n t] \tag{12.51}$$

Example 12.3

Determine the natural frequencies of free transverse vibration of a uniform cantilever beam as shown in Figure 12.7.

SOLUTION

Considering the fixed end of the cantilever at $x = 0$, the governing boundary conditions are

$$Y_n(0) = 0 \qquad \frac{d Y_n(0)}{dx} = 0$$

and at the free end $(x = 1)$

$$\frac{d^2 Y_n(l)}{dx^2} = 0 \qquad \frac{d^3 Y_n(l)}{dx^3} = 0$$

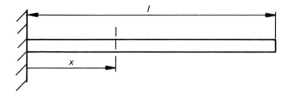

Figure 12.7

From these conditions it follows that

$$Y_n(0) = C_n \sin(0) + D_n \cos(0) + E_n \sinh(0) + F_n \cosh(0) = 0$$

$$\frac{dY_n(0)}{dx} = k_n[C_n \cos(0) - D_n \sin(0) + E_n \cosh(0) + F_n \sinh(0)] = 0$$

from which we obtain the relationships

$$D_n + F_n = 0 \quad \text{or} \quad D_n = -F_n$$

and

$$C_n + E_n = 0 \quad \text{or} \quad C_n = -E_n$$

We have, in addition,

$$\frac{d^2Y_n(l)}{dx^2} = -k_n^2[C_n \sin(k_n l) + D_n \cos(k_n l) - E_n \sinh(k_n l) - F_n \cosh(k_n l)] = 0$$

Substituting $D_n = -F_n$ and $C_n = -E_n$ gives

$$C_n(\sin k_n l + \sinh k_n l) + D_n(\cos k_n l + \cosh k_n l) = 0 \tag{12.52}$$

and

$$\frac{d^3Y_n(l)}{dx^3} = k_n^3(C_n \cos k_n l - D_n \sin k_n l - E_n \cosh k_n l - F_n \sinh k_n l) = 0$$

which after substitution of $D_n = -F_n$ and $C_n = -E_n$ gives

$$C_n(\cos k_n l + \cosh k_n l) + D_n(\sinh k_n l - \sin k_n l) = 0 \tag{12.53}$$

Writing equations 12.52 and 12.53 in matrix form we have

$$\begin{bmatrix} (\sin k_n l + \sinh k_n l) & (\cos k_n l + \cosh k_n l) \\ \hline (\cos k_n l + \cosh k_n l) & (\sinh k_n l - \sin k_n l) \end{bmatrix} \begin{Bmatrix} C_n \\ D_n \end{Bmatrix} = \begin{Bmatrix} 0 \\ 0 \end{Bmatrix} \tag{12.54}$$

The non-trivial solution of equation 12.54 is obtained by equating the determinant to zero, i.e.

$$\begin{bmatrix} (\sin k_n l + \sinh k_n l) & (\cos k_n l + \cosh k_n l) \\ (\cos k_n l + \cosh k_n l) & (\sinh k_n l - \sin k_n l) \end{bmatrix}_{\text{det}} = 0 \tag{12.55}$$

Expanding the determinant and utilizing the standard relationships $\cos^2 k_n l + \sin^2 k_n l = 1$, and $\cosh^2 k_n l - \sinh^2 k_n l = 1$ produces the equation

$$2 + 2\cos k_n l \cosh k_n l = 0 \quad \text{i.e.} \quad \cos k_n l \cosh k_n l = -1 \tag{12.56}$$

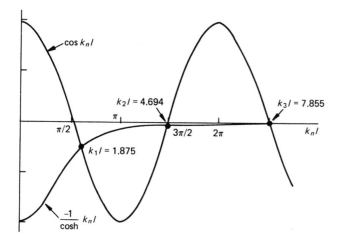

Figure 12.8

The transcendental equation 12.56 may be expressed in the form

$$\cos k_n l = -1/\cosh k_n l \qquad (12.57)$$

A first approximation to the roots of equation 12.57 can be obtained by plotting $\cos k_n l$ and $-1/\cosh k_n l$ to the common base of $k_n l$ (see Figure 12.8). Subsequently these approximations can be used in a process of iteration of equation 12.57 to produce the first five roots:

$$k_1 l = 1.875 \qquad k_2 l = 4.694 \qquad k_3 l = 7.855 \qquad k_4 l = 10.996 \qquad k_5 l = 14.137$$

Corresponding values for higher modes may be obtained from the equation

$$k_n l = \frac{(2n-1)\pi}{2} \qquad (12.58)$$

From the roots of equation 12.56 the relevant natural frequency of transverse vibration may be obtained from

$$\omega_n = k_n^2 \sqrt{\left(\frac{EJ}{\rho A}\right)} \qquad (12.59)$$

Mode shape
Since the matrix equation 12.54 is singular (i.e. its determinant is equal to zero), expansion results in a single independent equation. It follows, therefore, that either equation 12.52 or equation 12.53 may be used to establish the relationship between the constants C_n and D_n and subsequently the mode shape associated with any particular natural frequency. Using equation 12.52 gives

$$D_n = -\frac{\sin k_n l + \sinh k_n l}{\cos k_n l + \cosh k_n l} C_n \qquad (12.60)$$

Thus, from equation 12.48, the mode shape for the nth mode may be expressed as

$$Y_n(x) = C_n \left[\sin k_n l - \sinh k_n l - \left(\frac{\sin k_n l + \sinh k_n l}{\cos k_n l + \cosh k_n l} \right) (\cos k_n x - \cosh k_n x) \right] \quad (12.61)$$

and the complete response of the beam with all possible modes being induced will be given once again by equation 12.51.

12.5.1 Effects of rotary inertia and shear deformation

The basic equation of motion, equation 12.40, previously derived for the transverse vibration of a prismatic beam specifically excluded the effects of rotary inertia and shear deformation and is, therefore, intrinsically inaccurate, although the degree of inaccuracy is very small for the lower modes of vibration. The inaccuracy follows from the omission of two aspects of vibration, namely:

(1) the kinetic energy associated with the rotation of each element of the beam between the positions of maximum vibratory deformation, and
(2) the additional deflection induced in the beam by the lateral shear forces which produce a skewing effect on each and every element of the beam.

Inclusion of the rotary inertia effect adds a further term to equation 12.40, which becomes

$$EJ \frac{\partial^4 y}{\partial x^4} - \rho J \frac{\partial^4 y}{\partial x^2 \partial t^2} + \rho A \frac{\partial^2 y}{\partial t^2} = 0 \quad (12.62)$$

whilst the added effect of shear deformation modifies the basic equation to give

$$EJ \frac{\partial^4 y}{\partial x^4} - \frac{\rho EJ}{ZG} \cdot \frac{\partial^4 y}{\partial x^2 \partial t^2} + \rho A \frac{\partial^2 y}{\partial t^2} = 0 \quad (12.63)$$

where Z is a constant dependent on the shape of the beam cross-section and, for a rectangular section, is approximately 0.85. Solutions to equations 12.62 and 12.63 lead to the derivation of natural frequencies which are a few per cent more accurate than those obtained using thin beam theory, i.e. equation 12.40. This enhanced accuracy is usually only of importance where the higher modes of transverse vibration are of interest, i.e. where the beam will form shorter beam lengths between nodal points.

In general the correction required to be made to the frequencies obtained using thin beam theory is greater for shear deformation than for rotary inertia.

12.5.2 Transverse vibration of a rotating beam

In Section 12.2 we examined the transverse vibration of a stretched wire which was considered as a beam of negligible flexural rigidity, i.e. the product EJ is negligible; the only restoring effect, responding to a lateral displacement, was the in-plane tension (s). Subsequently we have extended our analysis to the transverse vibration of beams of finite flexural rigidity in the absence of in-plane tension. We can now combine both restoring effects and proceed to consider the transverse vibration of beams when both of these elements act simultaneously.

If we denote the flexural stiffness by K_f and consider the restraining effect of the in-plane tension as an analogous stiffness, generally referred to as the membrane (or geometric) stiffness K_g, then the total lateral stiffness, K_t, of the beam may be written as

$$(K_t)_n = (K_f)_n + (K_g)_n \tag{12.64}$$

Equation 12.64 implies that the individual stiffnesses of the beam are in parallel. If we consider equation 12.64 in relation to a single mass/elastic system, the natural frequencies of such a system may be obtained from the usual relationship, namely

$$\omega_n^2 = (K_t/M)_n \tag{12.65}$$

where M is the total equivalent mass of the system. Substituting for K_t in equation 12.65 gives

$$\omega_n^2 = (K_f/M)_n + (K_g/M)_n \tag{12.66}$$

or

$$\omega_n^2 = (\omega_n^f)^2 + (\omega_n)^2 \tag{12.67}$$

where ω_n^f is the flexural component of the natural frequency which may be determined in the normal way, i.e. from exact analysis or Rayleigh or Ritz energy methods, as covered in subsequent sections, and ω_n^g is the membrane component of the natural frequency. Equation 12.67 is a statement of an approximation first established by H. Lamb and R.V. Southwell ('The vibration of a spinning disc', *Proc. R. Soc. (London)*, *Ser. A*, **99**, 1921, 272–280) which showed that derivation of the natural frequency of the system by separate analysis of the flexural and membrane components, with subsequent addition as shown in equation 12.67, resulted in a value of natural frequency which represented a *lower bound* to the *true* value, i.e. the value that would be obtained if both components were included simultaneously in an *exact analysis*. Let us now devote our attention to the membrane component of the natural frequency and examine the factors that influence this value.

Consider the case of a thin prismatic beam, of cross-section A and length l, which is hinged at a central hub of negligible radius, and rotating at a uniform speed, Ω, as shown in Figure 12.9. (As a first approximation this may represent an aircraft propeller or helicopter rotor blade.)

If we consider an element of length δx, located at distance x from the hub, the corresponding stress distribution associated with this element will be as shown in Figure 12.10.

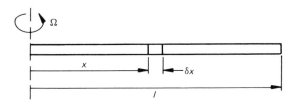

Figure 12.9

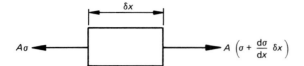

Figure 12.10

Thus for a uniform cross-section beam the elemental force acting on the element will be

$$\delta F = A\left(\sigma + \frac{d\sigma}{dx}\,\delta x\right) - A\sigma = A\,\delta x\frac{d\sigma}{dx} \qquad (12.68)$$

which will, of course, be equal and opposite to the centrifugal force acting on the element at that position on the beam, i.e.

$$\delta F = A\,\delta x\frac{d\sigma}{dx} = -\rho A\,\delta x\Omega^2 x \qquad (12.69)$$

giving

$$\frac{d\sigma}{dx} = -\rho\Omega^2 x \qquad (12.70)$$

where ρ is the material density of the beam. From Hooke's Law we have the relationship

$$\sigma = E\varepsilon = E\frac{du}{dx}$$

where E is the modulus of elasticity, ε the strain at section x, and u the longitudinal displacement at this point. From this we obtain

$$\frac{d\sigma}{dx} = E\frac{d^2u}{dx^2}$$

Substituting for $d\sigma/dx$ in equation 12.70 gives

$$E\frac{d^2u}{dx^2} = -\rho\Omega^2 x \qquad (12.71)$$

and integrating with respect to x produces

$$E\frac{du}{dx} = \frac{-\rho\Omega^2 x^2}{2} + B \qquad (12.72)$$

Integrating once again gives

$$Eu = \frac{-\rho\Omega^2 x^3}{6} + Bx + C \qquad (12.73)$$

B and C are constants which may be obtained from the governing boundary conditions in the usual manner, i.e. at $x = 0$, $u = 0$ which after substitution in equation 12.73 gives $C = 0$. Also, at $x = l$, $\varepsilon = du/dx = 0$. Substitution in equation 12.72 gives $B = \rho\Omega^2 l^2/2$.

The final form of equation 12.72 may now be expressed as

$$\sigma = E\frac{du}{dx} = \frac{\rho\Omega^2}{2}(l^2 - x^2) = \frac{\rho\Omega^2 l^2}{2}(1 - \eta^2) \tag{12.74}$$

where $\eta = x/l$. Now let us consider the beam to be performing a normal mode of vibration in the lateral direction at some natural frequency ω_n^g, giving rise to a lateral displacement

$$y = y(x, t) = Y_n(x)\,\phi(t) \tag{12.75}$$

where once again $Y_n(x)$ is the normal function depicting the mode on shape of the deflected form and $\phi(t)$ gives the temporal variation. For convenience of solution we shall use complex vector notation to define the deflection, i.e.

$$y(x, t) = \text{real part of vector } Y_n(x)\exp(i\omega_n^g t)$$

During the vibratory cycle the position and stress variation of the element δx will be as shown in Figure 12.11. The resulting net force δF in the vertical direction is therefore given by the expression

$$\delta F = A\left(\sigma + \frac{d\sigma}{dx}\delta x\right)\left[\frac{\partial y}{\partial x} + \frac{\partial}{\partial x}\left(\frac{\partial y}{\partial x}\right)\delta x\right] - A\sigma\frac{\partial y}{\partial x} \tag{12.76}$$

which, after neglecting the product of small quantities, gives

$$\delta F = A\,\delta x\left(\sigma\frac{\partial^2 y}{\partial x^2} + \frac{d\sigma}{dx}\cdot\frac{\partial y}{\partial x}\right) \tag{12.77}$$

Applying Newton's Second Law to the element gives

$$\delta F = \frac{\partial}{\partial t}\left(\rho A\,\delta x\frac{\partial y}{\partial t}\right)$$

and substituting for δF from equation 12.77 yields the relationship

$$\sigma\frac{\partial^2 y}{\partial x^2} + \frac{d\sigma}{dx}\cdot\frac{\partial y}{\partial x} = \rho\frac{\partial^2 y}{\partial t^2} \tag{12.78}$$

If we now replace y by its vector representation, i.e. $y = Re\,[Y_n(x)\exp(i\omega_n^g t)]$ we obtain

$$\sigma\frac{d^2 Y_n(x)}{dx^2} + \frac{d\sigma}{dx}\frac{d Y_n(x)}{dx} + \rho(\omega_n^g)^2\,Y_n(x) = 0 \tag{12.79}$$

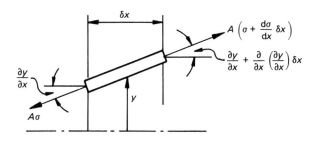

Figure 12.11

Substituting for σ and its derivative from equation 12.74 gives

$$(1 - \eta^2)\frac{d^2 Y_n(\eta)}{d\eta^2} - 2\eta \frac{d Y_n(\eta)}{d\eta} + \lambda_n Y_n(\eta) = 0 \qquad (12.80)$$

where

$$\lambda_n = 2\left(\frac{\omega_n^g}{\Omega}\right)^2$$

If $\lambda_n = n(n + 1)$, where n is a real number, then equation 12.80 is immediately recognized as Legendre's equation (named after Adrien Marie Legendre (1752–1833), a French mathematician noted for his work in the theory of numbers), which for any particular value of n has a solution termed the Legendre function (or polynomial), P_n, equal to $Y_n(\eta)$, for example

$$
\begin{aligned}
n = 0 \qquad & \lambda_n = 1 \qquad && P_0 = Y_0(\eta) = 1 \\
n = 1 \qquad & \lambda_n = 2 \qquad && P_1 = Y_1(\eta) = \eta \\
n = 2 \qquad & \lambda_n = 6 \qquad && P_2 = Y_2(\eta) = 0.5(3\eta^2 - 1) \\
n = 3 \qquad & \lambda_n = 12 \qquad && P_3 = Y_3(\eta) = 0.5(5\eta^3 - 3\eta) \\
n = 4 \qquad & \lambda_n = 15 \qquad && P_4 = Y_4(\eta) = 0.125(35\eta^4 - 30\eta^2 + 3) \\
n = 5 \qquad & \lambda_n = 30 \qquad && P_5 = Y_5(\eta) = 0.125(63\eta^5 - 70\eta^3 + 15\eta)
\end{aligned}
$$

or graphically as shown in Figure 12.12

For the particular problem being analysed an essential boundary condition is that the lateral displacement of the beam at the hub must be zero at all times, i.e. $Y_n(0) = 0$. It follows, therefore, that the only admissible values of P_n which satisfy this requirement

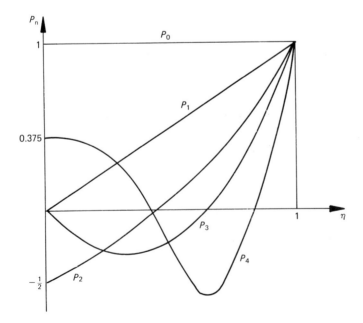

Figure 12.12

are P_1, P_3, P_5, etc., which correspond to the first, second and third modes of transverse vibration of the beam, respectively.

Substitution of the values for P_n subsequently yields the associated natural frequency equation, i.e.

for 1st mode, $n = 1$ giving

$$\lambda_1 = 2 = 2\left(\frac{\omega_1^g}{\Omega}\right)^2 \qquad \text{or} \qquad \omega_1^g = \Omega$$

for 2nd mode, $n = 3$ giving

$$\lambda_3 = 12 \qquad \text{and} \qquad \omega_3^g = \sqrt{6\Omega} = 2.45\Omega$$

and

for 3rd mode, $n = 5$ giving

$$\lambda_5 = 30 \qquad \text{and} \qquad \omega_5^g = \sqrt{15\Omega} = 3.87\Omega$$

Reverting to equation 12.67 we can now plot the variation of the natural frequency of the beam with change of rotational velocity as shown in Figure 12.13.

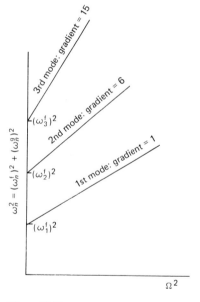

Figure 12.13

Example 12.4

Determine the first three natural frequencies and corresponding modes of free lateral vibration of a uniform steel cantilever rotating about an axis perpendicular to the length of the rod. Obtain these values for rotational speeds of zero, 1000 rev/min and 2000 rev/min.

Data:

- $E = 200 \times 10^9 \, \text{N/m}^2$.
- $\rho = 7600 \, \text{kg/m}^3$.
- $J = 65 \times 10^{-9} \, \text{m}^4$.
- $A = 1250 \times 10^{-6} \, \text{m}^2$.
- $l = 1 \, \text{m}$.

SOLUTION

(a) Flexural components for the first three modes of vibration are given by the solution to equation 12.58, i.e.

$$k_1 l = 1.875 \qquad k_2 l = 4.694 \qquad k_3 l = 7.855$$

where, in this case,

$$k_n = \left[\frac{\rho A}{EJ} (\omega_n^f)^2 \right]^{1/4} \qquad \text{or} \qquad \omega_n^f = k_n^2 \sqrt{\frac{EJ}{\rho A}}$$

Substituting for all constants gives

$$\omega_1^f = 130 \, \text{rad/s} = 20.7 \, \text{Hz}$$
$$\omega_2^f = 815 \, \text{rad/s} = 129.7 \, \text{Hz}$$
$$\omega_3^f = 2242 \, \text{rad/s} = 356.8 \, \text{Hz}$$

(b) The corresponding membrane components of the natural frequencies will be given by the expressions

$$\omega_1^g = \Omega \qquad \omega_3^g = 2.45\Omega \qquad \omega_5^g = 3.87\Omega$$

and the complete frequencies may be derived from

$$\omega_n^2 = (\omega_n^f)^2 + (\omega_{2n-1}^g)^2 \tag{12.81}$$

These are expressed in Table 12.1 for the three speeds quoted.

Table 12.1 Natural frequencies (Hz) for a rotating cantilever

		Rotational speed (rev/min)		
No.	Mode shape	0	1000	2000
1		20.7	26.6	39.2
2		129.7	136.0	153.2
3		356.8	363.7	379.5

12.6   Energy methods

There are many instances concerning the transverse vibration of distributed mass/elastic systems where it is not always possible to obtain an exact solution to the problem. In other cases, where an exact solution does exist, extraction of even the fundamental mode may be somewhat laborious and as such it is often less tedious to adopt an approximate solution. The degree of error introduced by using an approximate method will depend on the form of the solution and the assumptions made regarding the mode of vibration of the beam, e.g. the more boundary conditions satisfied the greater the degree of accuracy achieved in determining the natural frequencies and associated modes of vibration. In many practical problems, however, it may be that only the fundamental mode of vibration of a beam is likely to lie within the operating range of the device to which it is attached (or of which it forms part). In such cases a fairly high level of accuracy can be achieved with the minimum amount of effort.

Although there are many forms of energy methods, we shall confine our analysis to two of the most commonly used versions, namely, the Rayleigh method (named after Lord Rayleigh, John William Strutt (1842–1919), English physicist, Nobel Prize winner 1904) and the Ritz method (or Rayleigh–Ritz) method, which come within the scope of what is generally referred to as 'energy methods'.

12.6.1   Rayleigh method

All of the distributed mass/elastic systems considered in this chapter have been assumed to be conservative systems, i.e. no energy dissipation has taken place. Consequently vibration of these systems represents an exchange (without loss) of energy between two energy stores, namely potential (or strain) and kinetic. At any instant during the vibratory cycle, the total energy of vibration of the system, whether it be a wire, beam or shaft, will be the sum of each of these contributions. As the beam or wire passes through its mean or equilibrium position, the energy will be wholly kinetic; conversely, at the maximum deflection position the energy will be wholly potential or strain.

In the Rayleigh method of approximation, the maximum values of kinetic energy (KE) and strain energy (SE) are equated and the natural frequency of vibration of the system for the mode (usually the fundamental) in question is thereby extracted.

Since both the KE and SE terms depend on the displacement of the beam or wire, it is necessary to know or assume what the deflected shape will be. Rayleigh has shown that the choice of the deflected form of the beam or wire does not greatly affect the accuracy of the solution provided the geometric boundary conditions are satisfied. Selecting a deflected form that does not correspond exactly to the true deflected form is equivalent to introducing additional constraints to the system, which increases the stiffness of the system with consequent increase in the strain energy for a given displacement. As a direct result of this, the natural frequencies of vibration of a system, determined by means of the Rayleigh method, are generally a few per cent higher than the true values, although if the assumed deflected form happens to coincide with the true deflected form the solution will be exact.

As an example of the use of the Rayleigh energy method and to assess the accuracy of this form of approximate solution, let us refer once again to the cantilever beam shown in Figure 12.7 but in this case, rather than determine the normal function $Y_n(x)$

as before, we shall assume a deflected shape for the fundamental mode corresponding to a quarter cosine wave for ease of differentiation and integration. Thus

$$y = y(x, t) = Y_n(x) \sin \omega_n t$$
$$= Y(x) \sin \omega t \qquad \text{for the fundamental mode} \qquad (12.82)$$

where

$$Y(x) = Y_0 \left(1 - \cos \frac{\pi x}{2l} \right) \qquad (12.83)$$

and Y_0 is the maximum deflection at the end of the beam, i.e. at $x = l$.

Differentiating equation 12.83 with respect to x gives

$$\frac{dY(x)}{dx} = Y_0 \frac{\pi}{2l} \sin \frac{\pi x}{2l} \qquad (12.84)$$

The geometric boundary conditions may now be checked, i.e.

$$Y(0) = Y_0 [1 - \cos(0)] = 0$$
$$X(l) = Y_0 [1 - \cos(\pi/2)] = Y_0$$

$$\frac{dY(0)}{dx} = Y_0 \frac{\pi}{2l} \sin(0) = 0$$

$$\frac{dY(l)}{dx} = Y_0 \frac{\pi}{2l} \sin \frac{\pi}{2} = \frac{Y_0 \pi}{2l} \neq 0$$

Thus the geometric (i.e. displacement and slope) boundary conditions are satisfied by the choice of deflection function. Consider now the energy content of the vibrating beam.

Strain energy (SE)
From simple beam theory, we have the relationship

$$SE = \frac{1}{2EJ} \int_0^l M^2(x) \, dx$$

where $M(x)(= EJ(\partial^2 y / \partial x^2))$ is the bending moment at section x. Substituting for $M(x)$ gives

$$SE = \frac{1}{2EJ} \int_0^l \left(EJ \frac{\partial^2 y}{\partial x^2} \right)^2 dx$$
$$= \frac{EJ}{2} \int_0^l \left[\frac{d^2 Y(x)}{dx^2} \right]^2 dx \sin^2 \omega t \qquad \text{for a prismatic beam}$$

and

$$\widehat{SE} = \frac{EJ}{2} \int_0^l \left[\frac{d^2 Y(x)}{dx^2} \right]^2 dx \qquad (12.85)$$

where $\widehat{SE}$ is the maximum value of SE during the vibratory cycle. Differentiating equation 12.83 twice with respect to x and substituting in equation 12.85 gives

$$\widehat{SE} = \frac{EJ}{32} \cdot \frac{Y_0^2 \pi^4}{l^4} \int_0^l \cos^2\left(\frac{\pi x}{2l}\right) dx$$

$$= \frac{EJ}{64} \cdot \frac{Y_0^2 \pi^4}{l^4} \int_0^l \left[1 + \cos\left(\frac{\pi x}{l}\right)\right] dx$$

$$= \frac{EJ}{64} \cdot \frac{Y_0^2 \pi^4}{l^3} \tag{12.86}$$

Kinetic energy (KE)
Consider an element of length δx, as shown in Figure 12.14, located at position x, where at any instant the lateral displacement of the beam is $y = y(x, t)$, and mass of element $= \rho A\, \delta x$ where ρ is the mass density of beam material and A is the cross-sectional area of the beam. It follows therefore that the kinetic energy, δKE, possessed by the element will be $\frac{1}{2}\rho A\, \delta x\, (\partial y/\partial t)^2$, i.e.

$$\delta KE = \tfrac{1}{2}\rho A\, \delta x\, [\omega\, Y(x) \cos \omega t]^2$$

and

$$\delta\widehat{KE} = \tfrac{1}{2}\rho A\, \delta x\, \omega^2 [Y(x)]^2 = \text{maximum kinetic energy possessed by the}$$
$$\text{element during vibration}$$

Correspondingly, for the complete beam we have

$$\widehat{KE} = \tfrac{1}{2}\rho A \omega^2 \int_0^l [Y(x)]^2\, dx$$

Substituting for $Y(x)$ from equation 12.83 gives

$$\widehat{KE} = \tfrac{1}{2}\rho A \omega^2 \int_0^l Y_0^2 \left[1 - \cos\left(\frac{\pi x}{2l}\right)\right]^2 dx \tag{12.87}$$

which after integrating produces

$$\widehat{KE} = \rho A l \omega^2 Y_0^2 \left[0.75 - \frac{2}{\pi}\right] \tag{12.88}$$

Equating the maximum values of strain and kinetic energy, i.e. equations 12.86 and

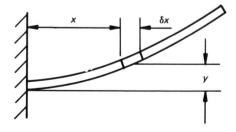

Figure 12.14

12.88 respectively, according to the requirements of the Rayleigh method, results in the expression

$$\omega = \frac{\pi^2}{l^2} \frac{\sqrt{(EJ/\rho A)}}{8\sqrt{(0.75 - 2/\pi)}} = \frac{3.6638}{l^2} \sqrt{\left(\frac{EJ}{\rho A}\right)} \tag{12.89}$$

From Example 12.3, the exact solution corresponding to the fundamental mode of lateral vibration of a prismatic cantilever beam was given by equation 12.59, i.e.

$$\omega_n = k_n^2 \sqrt{\left(\frac{EJ}{\rho A}\right)}$$

where, for the fundamental mode, $k_n = k_1 = 1.875/l$, giving

$$\omega_1 = \frac{3.5156}{l^2} \sqrt{\left(\frac{EJ}{\rho A}\right)} \tag{12.90}$$

Consequently the value for the frequency corresponding to the fundamental mode of vibration of the cantilever beam, obtained by the Rayleigh energy method, is seen to be 4.2 per cent greater than the true value. The reason for this error becomes apparent when we further examine the compliance of the assumed normal function, $Y(x) = Y_0[1 - \cos(\pi x/2l)]$, with the boundary conditions at the free end of the beam. At the free end we know that both the bending moment $EJ[d^2 Y(x)/dx^2]$ and the shear force $EJ[d^3 Y(x)/dx^3]$ *should* be zero, therefore when we differentiate $Y(x)$ with respect to x twice and three times and then substitute l for x, we find that

$$\frac{d^2 Y(x)}{dx^2} = 0 \qquad \text{when } x = l$$

(i.e. compliance with boundary condition)

$$\frac{d^3 Y(x)}{dx^3} = -\frac{\pi^3 Y_0}{8l^3} \qquad \text{when } x = l$$

Therefore the assumed normal function implies that a shear force exists at the free end of the beam and it is this extra constraint on the system that gives rise to the 4.2 per cent upper bound to the exact solution.

12.6.2 Ritz method

As was seen in Section 12.6.1, the Rayleigh energy method is useful in the determination of the natural frequency corresponding to the fundamental mode of vibration of a continuous elastic system. By equating the maximum values of strain and kinetic energy, the relevant frequency was obtained from the equation

$$\omega^2 = \frac{EJ}{\rho A} \frac{\displaystyle\int_0^l \left[\frac{d^2 Y(x)}{dx^2}\right]^2 dx}{\displaystyle\int_0^l [Y(x)]^2 dx} \tag{12.91}$$

which is sometimes referred to as the 'Rayleigh energy quotient'. The degree of accuracy achieved by this method can be improved greatly by the use of the Ritz method of approximate solution and this approach may also be used to obtain the frequencies corresponding to the higher modes of vibration of the system.

In using the Ritz method, an assumed deflection form (normal function), $Y(x)$, representing the mode of vibration, is considered jointly with several parameters, the magnitude of which are used in such a manner as to reduce to a minimum the frequency of vibration being considered. Thus if we let $\phi_1(x)$, $\phi_2(x)$, $\phi_3(x),\ldots,\phi_n(x)$ be a series of functions, each satisfying the governing boundary conditions (or as many as possible) and also suitable for representing $Y(x)$, we can write

$$Y(x) = a_1\phi_1(x) + a_2\phi_2(x) + a_3\phi_3(x) + \cdots + a_n\phi_n(x) \tag{12.92}$$

where $a_1, a_2, a_3,\ldots,a_n$ are *minimizing coefficients* which are used to reduce the right-hand side of equation 12.91 to a minimum. Consequently a series of equations of the form

$$\frac{\partial}{\partial a_n}\frac{\displaystyle\int_0^l\left[\frac{d^2Y(x)}{dx^2}\right]^2 dx}{\displaystyle\int_0^l [Y(x)]^2 dx} = 0 \tag{12.93}$$

may be derived.

Performing the differentiation in the usual manner and equating the numerator of such to zero gives

$$\int_0^l [Y(x)]^2\,dx\,\frac{\partial}{\partial a_n}\left\{\int_0^l\left[\frac{d^2Y(x)}{dx^2}\right]^2 dx\right\} - \int_0^l\left[\frac{d^2Y(x)}{dx^2}\right]^2 dx\cdot\frac{\partial}{\partial a_n}\left\{[Y(x)]^2\,dx\right\} = 0 \tag{12.94}$$

Now from equation 12.91 we have the relationship

$$\int_0^l\left[\frac{d^2Y(x)}{dx^2}\right]^2 dx = \frac{\omega^2\rho A}{EJ}\int_0^l [Y(x)]^2\,dx$$

which, after substitution in equation 12.94, gives

$$\frac{\partial}{\partial a_n}\int_0^l\left\{\left[\frac{d^2Y(x)}{dx^2}\right]^2 - \frac{\rho A\omega^2}{EJ}[Y(x)]^2\right\}dx = 0 \tag{12.95}$$

Thus a series of homogeneous equations, linear in $a_1, a_2, a_3,\ldots,a_n$, is obtained, the number of which will be equal to the number of minimizing coefficients used in equation 12.92. This system of equations will produce, for the constants a_1, a_2, $a_3,\ldots,a_n$, solutions that are non-trivial, only if the determinant of the equations is zero. This results in the formation of a frequency equation from which the frequencies corresponding to the number of modes of vibration being considered may be obtained. For example if equation 12.92 contains three terms, this is equivalent to reducing the continuous distributed system to a three degree of freedom system having three natural frequencies.

The variation of choice of normal function and the procedure adopted for the calculation of consecutive frequencies is indicated in the following example, which, for comparison with the previously considered exact and Rayleigh energy solutions, relates to a uniform section cantilever beam.

Example 12.5

Consider the expression for the deflected form of a uniform cantilever beam, performing lateral vibration at a frequency $\omega\,\mathrm{rad/s}$, to be

$$y = y(x, t) = (a_1x^2 + a_2x^3 + a_3x^4 + \cdots + a_nx^{n+1})\sin \omega t = Y(x)\sin \omega t \qquad (12.96)$$

Each term must, of course, satisfy the geometric boundary conditions which in this case are

$$y(0, t) = \frac{\partial y(0, t)}{\partial x} = 0 \qquad (12.97)$$

which implies

$$Y(x) = \frac{\mathrm{d}\,Y(x)}{\mathrm{d}x} = 0 \qquad \text{at } x = 0 \qquad (12.98)$$

From equation 12.96 for the nth term we have

$$Y(x) = a_nx^{n+1} \qquad (12.99)$$

and

$$\mathrm{d}Y(x)/\mathrm{d}x = (n + 1)a_nx^n \qquad (12.100)$$

Thus, for all values of n, equation 12.98 is satisfied. To study the rate of convergence of the Ritz solution, let us vary, sequentially, the number of terms used to define the deflection function, i.e. equation 12.96.

SINGLE-TERM FUNCTION

i.e.

$$Y(x) = a_1x^2 \qquad (12.101)$$

also

$$[Y(x)]^2 = a_1^2x^4 \qquad [\mathrm{d}^2 Y(x)/\mathrm{d}x^2]^2 = 4a_1^2$$

Substituting for these expressions in equation 12.95 gives

$$\frac{\partial}{\partial a_1}\left[\int_0^l \left(4a_1^2 - \frac{\omega^2\rho A}{EJ}a_1^2x^4\right)\mathrm{d}x\right] = 0$$

which after integrating gives

$$\frac{\partial}{\partial a_1}\left(4a_1^2l - \frac{\omega^2\rho A}{5EJ}a_1^2l^5\right) = 0$$

Subsequent differentiation yields

$$2a_1\left(4l - \frac{\omega^2\rho A}{5EJ}l^5\right) = 0$$

which produces the non-trivial solution

$$\omega = \frac{\sqrt{20}}{l^2}\sqrt{\left(\frac{EJ}{\rho A}\right)} = \frac{4.472}{l^2}\sqrt{\left(\frac{EJ}{\rho A}\right)} \qquad (12.102)$$

When compared with the corresponding exact natural frequency, i.e.

$$\omega = \frac{3.5156}{l^2} \cdot \frac{EJ}{\rho A}$$

the Ritz solution is seen to be in error by $+27.2$ per cent.

TWO-TERM FUNCTION

i.e.

$$Y(x) = a_1 x^2 + a_2 x^3 \tag{12.103}$$

Substitution of equation 12.103 and its derivative into equation 12.95, followed by integration and differentiation with respect to a_1 and a_2 sequentially, produces the equations

$$a_1(8l - 0.4\lambda l^5) + a_2\left(12l^2 - \frac{\lambda l^6}{3}\right) = 0 \tag{12.104}$$

and

$$a_1\left(12l^2 - \frac{\lambda l^6}{3}\right) + a_2\left(24l^3 - \frac{2}{7}\lambda l^7\right) = 0 \tag{12.105}$$

where, for convenience, constant

$$\lambda = \omega^2 \rho A / EJ \tag{12.106}$$

Rewriting in matrix form, we have

$$\begin{bmatrix} 8l^2 - 0.4\lambda l^5 & 12l^2 - \dfrac{\lambda l^6}{3} \\ 12l^2 - \dfrac{\lambda l^6}{3} & 24l^3 - \dfrac{2}{7}\lambda l^7 \end{bmatrix} \begin{Bmatrix} a_1 \\ a_2 \end{Bmatrix} = \begin{Bmatrix} 0 \\ 0 \end{Bmatrix} \tag{12.107}$$

Equating the determinant of equation (12.107) to zero gives the frequency equation

$$\frac{\lambda^2 l^8}{315} - \frac{136\lambda l^4}{35} + 48 = 0 \tag{12.108}$$

which has two roots, namely

$$\lambda_1 = 12.4799/l^4 \qquad \text{and} \qquad \lambda_2 = 1211.52/l^4$$

Substitution of these values into equation 12.106 yields the natural frequencies

$$\omega_1 = \frac{3.5327}{l^2} \sqrt{\left(\frac{EJ}{\rho A}\right)} \tag{12.109}$$

and

$$\omega_2 = \frac{34.8069}{l^2} \sqrt{\left(\frac{EJ}{\rho A}\right)} \tag{12.110}$$

In this solution ω_1 is the second approximation to the fundamental frequency with the error being reduced to $+0.48$ per cent compared with the true values; ω_2 is a first approximation of the natural frequency corresponding to the second mode of vibration of the beam. Comparison of the relevant true natural frequency, given by equation 12.59, with $k_2 = 4.694/l$, i.e. $\omega_2 = (22.0336/l^2)\sqrt{(EJ/\rho A)}$, indicates an error of magnitude equal to $+58$ per cent.

By increasing the number of terms in the normal function to 3, the error in the estimated value for ω_1 will be reduced to less than $+0.1$ per cent and that for ω_2 reduced to $+8.5$ per cent.

In addition a first approximation for the natural frequency associated with the third mode of vibration will be obtained.

The Ritz method is seen, therefore, to converge fairly rapidly to values of natural frequency which are within the levels of acceptance for engineering purposes. The limiting factor in the choice of the number of terms used to define the deflected form of the beam is the complexity of the differentiation and integration equation 12.95.

12.7 Whirling of shafts

Throughout this chapter we have been concerned with the vibration of essentially stationary systems, apart from Section 12.5.2, where centrifugal effects on a rotating cantilever beam were examined. Of equal importance, however, to designers and operators of rotodynamic machinery, is another phenomenon, which, although not strictly a vibration, is nevertheless the source of similar problems, namely **shaft whirling**. This condition is induced in rotating shafts, in particular thin flexible shafts, as a result of the centre of mass of the shaft section not coinciding exactly with the geometric axis of the shaft, which can arise due to inhomogeneity of the shaft material and the finite limits of tolerance applied during manufacture. If the rotational speed of such a shaft is gradually increased, eventually a speed will be reached at which violent instability of the shaft is encountered; this instability manifests itself in the form of the shaft deflecting into a single bow, with the shaft apparently hinged at the bearings and whirling around in a similar fashion to a skipping rope.

If this condition is allowed to continue, by maintaining the rotational speed at its initial value, the deflection of the shaft is likely to increase to such a level that failure may eventually occur; however, if the speed at which the instability is induced is quickly run through, the deflection of the shaft will diminish and the shaft will rotate once again in a stable manner, until a higher speed is encountered at which point further instability will appear, inducing a deflected form for the shaft corresponding to a double-bow mode, and so on. The rotational speed of the shaft corresponding to these conditions of instability are known as **critical speeds of whirling**, and may be present, not only on unloaded shafts, but also on shafts carrying concentrated leads such as rotors.

Let us consider the simplest case where a heavy rotor of mass M is attached to a thin shaft of negligible mass, as shown in Figure 12.15.

Let BB be the line between the bearing centres, O the geometric centre of the shaft and G the centre of gravity or mass of the rotor. As previously commented upon, the centre of mass G and the geometric centre of the shaft. Let e be the distance between O and G, i.e. the eccentricity, and r the deflection of the shaft at the rotor position. If we ignore gravitational effects there are, consequently, two forces acting

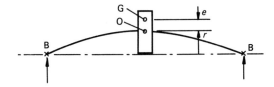

Figure 12.15

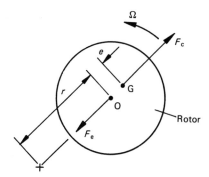

Figure 12.16

on the rotor, namely

(1)   the elastic restoring force F_e offered by the deflected shaft, which tends to return the shaft to the equilibrium or unconstrained position, i.e. tends to straighten the shaft, and

(2)   the centrifugal force F_c produced by the rotation of the shaft and acting at G.

The former force will be the product of the lateral bending stiffness, K, of the shaft and the radial deflection, r, of the shaft; the latter force will be the product of the rotor mass, the square of the rotational speed and the distance of the centre of mass from the line of the bearings. The line of action of each of these forces is as shown in Figure 12.16.

Thus, assuming that G lies out-board of O, we have

$$F_e = Kr \qquad F_c = M\Omega^2(r + e)$$

Equating these for equilibrium of the rotor gives

$$Kr = M\Omega^2(r + e) \tag{12.111}$$

or

$$Kr - M\Omega^2 r = M\Omega^2 e \tag{12.112}$$

and rearranging gives

$$r = \frac{M\Omega^2 e}{K - \Omega^2 M} = \frac{e\Omega^2}{K/M - \Omega^2} \tag{12.113}$$

From Chapter 9 we know that for a single degree of freedom system $K/M = \omega^2$ where ω is the undamped natural frequency of transverse vibration of the rotor on the shaft. Substituting in equation 12.113 gives

$$r = \frac{ef^2}{1 - f^2} \qquad \text{where } f = \Omega/\omega \tag{12.114}$$

Equation 12.114 is similar to that for the response of a single degree of freedom undamped system acted upon by a force proportional to the square of the rotational speed of the shaft. The graphical representation of equation 12.114 is shown in Figure 12.17.

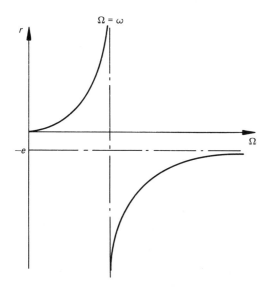

Figure 12.17

At $f = 1$ the deflection of the shaft builds up to extremely large levels and would become infinite if it were not for the finite degree of damping which is present in all practical systems (i.e. hysteretic, viscous, aerodynamic). This condition represents the whirling or critical speed, Ω_c, of the system. Consequently Ω_c may be obtained from the force equation 12.111, i.e. $Kr = M\Omega_c^2(r + e)$ or

$$\Omega_c^2 = \frac{Kr}{M(r + e)} = \frac{K}{M(1 + e/r)}$$

and at a critical whirling speed, $r \to \infty$, giving $e/r \to 0$, from which we obtain

$$\Omega_c = \sqrt{(K/M)} \tag{12.115}$$

which is the same as the natural frequency of transverse vibration of the shaft/rotor system.

It should be noted that when the running speed, Ω, is less than the critical speed, Ω_c, r and e have the same sign, i.e. the centre of mass G lies further from line BB than does O as shown in Figure 12.15; however, when Ω is greater than Ω_c, r, and e are of opposite sign which implies that G then lies between BB and O, as shown in Figure 12.18. As Ω is further increased beyond the critical condition, r approaches the value $-e$, i.e. G tends to a position coinciding with BB and the rotor rotates about its centre of mass with a high degree of stability.

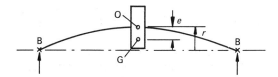

Figure 12.18

(Note that the foregoing analysis was conducted on the assumption that the bearings at BB were completely rigid. In many practical situations, particularly with pedestal bearings, this will certainly not be the case, and the derivation of the critical whirling speeds of the shaft/rotor system will require a more rigorous study, which is beyond the scope of this text.)

Problems

12.1 A uniform bar AB of length l and circular cross-section is supported freely in bearings. A harmonic axial torque $T_0 \cos \Omega t$ is applied to the end A, T_0 and Ω being constants. The material of the bar has density ρ and the torsional rigidity is GJ_0.
  Given that

$$\frac{\partial^2 \theta}{\partial x^2} = \frac{\rho}{G} \frac{\partial^2 \theta}{\partial t^2}$$

where θ is the angular displacement at any section x and at time t, show that the amplitude of steady-state forced torsional oscillations at the end B is given by

$$\frac{T_0 \operatorname{cosec} \lambda l}{\lambda GJ_0} \qquad \text{where } \lambda^2 = \Omega^2 \rho / G$$

12.2 Prove that the equation governing the free transverse vibration, $y(x, t)$, of stretched wire is

$$\frac{\partial^2 y}{\partial x^2} = \frac{1}{c^2} \cdot \frac{\partial^2 y}{\partial t^2} \qquad y = y(x, t)$$

where $c = \sqrt{(s/m)}$, s is the constant tension, and m is the mass per length. Assume that

$$y(x, t) = Y_n(x)[A_n \cos \omega_n t + B_n \sin \omega_n t]$$

where A_n and B_n are constant and $Y_n(x)$ is the normal function for the nth mode at frequency ω_n. Show that, for a stretched wire rigidly held at both ends,

$$\omega_n = \frac{n\pi}{l}\sqrt{\frac{S}{m}} \qquad n = 1, 2, 3 \dots$$

where l is the length of the wire.

A boiler tube, rigidly held at both ends, was found to have a fundamental undamped natural frequency of transverse vibration of 75 Hz when tested in a laboratory at normal room temperature. Under normal operating conditions, however, the tube is at a constant uniform temperature of 100°C above room temperature.

Given that the length and mass density of the tube are 3 m and 7600 kg/m³ respectively, estimate the value of the fundamental frequency under these conditions. Assume that Young's modulus $E = 200\,\text{GN/m}^2$ and α (the coefficient of linear thermal expansion) $= 1.2 \times 10^{-5}$ per °C. Assume that in-plane thermal stress $= -E\alpha T$, where T is the temperature above ambient.

Answer: 69 Hz.

12.3 List the important points relating to the Rayleigh energy method as applied to the free transverse vibration of beams. Figure 12.19 shows a beam of uniform flexural rigidity, EJ, and mass per length, m, rigidly supported at one end by the elastic foundation, of stiffness k, at the other end. At mid-span, the beam carries a circular mass of value M and diametrical mass moment of inertia I.

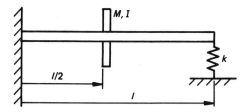

Figure 12.19

Assuming a fundamental vibratory normal function of the form

$$y(x) = A_0\left(1 - \cos\frac{\pi x}{2l}\right)$$

establish an approximate expression for the fundamental natural frequency of transverse vibration of the beam.

Answer: $\omega^2 = \left[\dfrac{EJ\pi^4}{64l^3} + \dfrac{k}{2}\right]\Bigg/\left[0.113\,38\,ml + 0.043M + \dfrac{0.25\pi^2 I}{4l^2}\right]$.

12.4 Show that for a beam of length l, encastré at both ends, the governing equation for the roots of k_n is

$$\cos k_n l \cosh k_n l = 1$$

where

$$k_n = \left(\frac{m\omega_n^2}{EJ}\right)^{1/4}$$

and EJ is the flexural rigidity and m the mass per length.

12.5 A beam of constant flexural rigidity EJ and mass per length m is built into a rigid support at one end and simply supported at the other end. Show that the natural frequencies of free transverse vibration of such a beam may be obtained from the solution of the equation

$$\tan(k_n l) = \tanh(k_n l) \tag{1}$$

In this equation, $k_n = (m\omega_n^2/EJ)^{1/4}$ where ω_n is the natural frequency associated with the nth mode of vibration, and l is the total length of the beam. Start your solution by assuming that the normal transverse vibratory function $Y_n(x)$ is

$$Y_n(x) = A_n(\cos k_n x + \cosh k_n x) + B_n(\cos k_n x - \cosh k_n x)$$
$$+ C_n(\sin k_n x + \sinh k_n x) + D_n(\sin k_n x - \sinh k_n x)$$

where A_n, B_n, C_n and D_n are constants for the nth mode.

Sketch, roughly, the mode shapes for values of $n = 1$, 2 and 3.

Demonstrate by means of a graphical representation of equation 1 that, for values of $n > 1$, approximate values for ω_n can be obtained from the expression

$$\omega_n = \frac{[(n - 0.75)\pi]^2}{l^2} \sqrt{\left(\frac{EJ}{m}\right)}$$

13
Introduction to vibratory control

13.1 Introduction

So far, in our studies relating to vibration, and in particular to forced vibration, we have mainly considered the displacement response of systems under the action of harmonic forcing. In Section 9.3.2, however, we investigated the amplitude of the subsequent force which would be transmitted to the system supports (frame force) as a result of the combined action of the harmonic force acting on the system and the consequential inertia of the vibrating mass. With reference to this investigation, the reader will be aware that for moderate degrees of damping, if the frequency of the applied force is equal, or close, to the undamped natural frequency of the system, then an extremely large amplitude harmonic frame force will prevail. In practice such a situation can be avoided by carefully designing the system in such a way as to ensure that none of its undamped natural frequencies is close to the frequency of any applied force(s) acting on the system under normal operating conditions. Very often, however, the dynamicist is confronted with a situation whereby after the system has been designed, manufactured and assembled, it is found that there is a resonance problem, i.e. the frequency of one (or more) of the applied forces is equal, or close, to one (or more) of the undamped natural frequencies of the system thus resulting in unacceptably high levels of vibratory response and frame forces. Also, due to financial and other design considerations, it is very seldom permissible either to redesign the system in order to effect changes in the natural frequencies, or to change the frequency or magnitude of the forces acting on the system. It is in situations such as this that the dynamicist has to select, design and implement some form of **vibratory control**, i.e. a means by which the vibration problem will be alleviated without having to alter the original system. Generally, vibratory control systems are classified under two distinct categories, namely **active** and **passive** systems, which will be explained in the remaining sections of this chapter. Furthermore, throughout this chapter we shall assume that damping is negligible.

13.2 Active vibratory control

This form of vibratory control is finding increasing popularity, especially in situations where the vibration has components at more than one frequency and/or where it is required to exercise vibratory control over a wide range of applied forcing frequencies. However, in an attempt to introduce it to the reader in its most basic form, let us consider the simple system shown in Figure 13.1 where a single degree of freedom spring/mass system is acted upon by the out-of-balance force as shown. Furthermore it will be assumed that the forcing frequency, Ω, is equal, or close, to the natural frequency of the system, $\omega = \sqrt{(K/M)}$, thus suggesting an unacceptably high magnitude of vibratory displacement and frame force at the foundations.

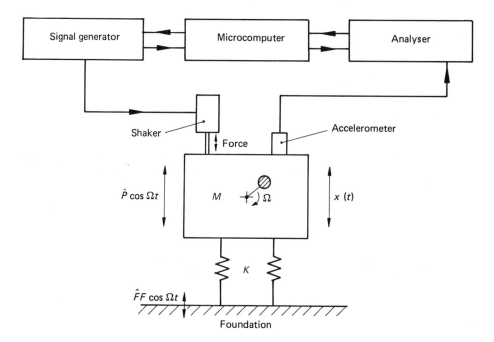

Figure 13.1

Let us now place on the mass a vibration transducer, normally an accelerometer, which will convert the vibration into an electrical signal of frequency equal to Ω and of an amplitude proportional to that of the peak vibratory acceleration of the mass. As shown in Figure 13.1 this signal is now relayed to a signal analyser/micro-computer/generator system, the function of which is to supply to the shaker a signal which will be converted into a second harmonic force acting on the mass, of the same magnitude as the force causing the vibration but 180° out of phase with it, thus suppressing the vibration. Let us now consider the stages by which this can be achieved.

(i) The signal received by the analyser from the accelerometer is stored as a complex function, $\vec{V}_1$, which describes the amplitude of the signal and its phase relative to some *constant datum pulse* generated and maintained within the analyser as shown in Figure 13.2, such that $\vec{V}_1 = A\underline{/\phi}$. Hence we can write

$$\vec{V}_1 = \vec{T}\vec{F} \tag{13.1}$$

where $\vec{F}$ is the complex function describing the unknown force acting on the mass to cause the vibration, and $\vec{T}$ represents the unknown transfer operator describing the relationship between this force and the vibration trace stored within the analyser.

(ii) Now let the generator produce some electric signal (voltage) $\vec{S}_1$, at the same

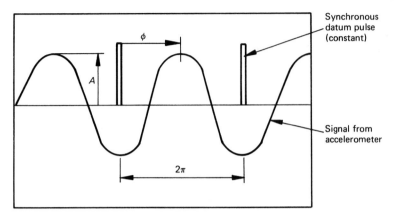

Figure 13.2

frequency as that of the vibration trace $\vec{V}_1$ and of some amplitude and phase relative to the datum pulse described above, the details of which are also stored in the analyser. Hence, the force produced by the shaker on the mass due to this generated signal can be written as $\vec{F}_t$, where

$$\vec{F}_t = \vec{G}\vec{S}_1 \tag{13.2}$$

and $\vec{G}$ is the unknown transfer operator describing the relationship between the force $\vec{F}_t$ and the generated signal $\vec{S}_1$. As a result of this action, the vibratory trace recorded and stored by the analyser will now change to $\vec{V}_2$ such that

$$\vec{V}_2 = \vec{T}(\vec{F} + \vec{G}\vec{S}_1) = \vec{T}\vec{F} + \vec{T}\vec{G}\vec{S}_1 \tag{13.3}$$

Hence subtracting equation 13.1 from equation 13.3 gives

$$\vec{T}\vec{G} = (\vec{V}_2 - \vec{V}_1)/\vec{S}_1 \tag{13.4}$$

and having solved for $\vec{T}\vec{G}$ the analyser can then proceed to compute and activate the necessary signal $\vec{S}_2$ which would reduce the vibration trace to zero, i.e. $\vec{V}_2 = 0$, in equation 13.4 such that

$$\vec{S}_2 = -\vec{V}_1/\vec{T}\vec{G} \tag{13.5}$$

The above can be described as a simple exercise implementing active vibratory control to a single degree of freedom system at one particular frequency. In practice, however, most systems have more than one degree of freedom with as many associated natural frequencies, all in their own way contributing to the overall vibratory problem. In such cases the active vibratory control system adopted will be more complex than that described above, although the basic principle remains the same, i.e. the vibration is eliminated or controlled by means of the application of secondary forcing to the system.

13.3   Passive vibratory control

In this case vibratory control is effected by superimposing on the original system devices that either *isolate* the vibrating system from the foundations on which it is mounted, or *absorb* the excessive vibration of the system.

With reference to *vibration isolation*, as we have seen in Section 9.3.2, the most common form is spring/damper devices, whereby, with suitably selected values of stiffness and damping coefficients, the amplitude of the harmonic frame force can be greatly reduced. This is the most common form of absorber used in automobiles, e.g. shock absorbers or MacPherson struts. More recently, however, two very interesting and effective isolators have been designed primarily for use in modern helicopters; these being the nodalized beam isolator (nodalized helicopter (description), *Aircraft Eng.* **45**(5), 1973, 15–16) and the dynamic anti-resonant vibration isolator (W.G. Flannelly, *Dynamic anti-resonant vibration isolator*, US Patent No. 3 322 379, Kaman Aircraft Corporation, 1967).

13.3.1   Nodalized beam isolator

This isolator, as shown in Figure 13.3, is based on the principle that a flexible beam, carrying a central mass requiring isolation, has two nodes (points of zero vibratory amplitude) when vibrating in the fundamental free–free lateral normal mode. Therefore, if the beam when vibrating at the natural frequency of this mode, which is equal to the forcing frequency, Ω, is supported at these two nodes, no vibration will be transmitted from the beam to the supports at the foundations. This is a simplification of the actual device, but it is used and effective isolation is claimed.

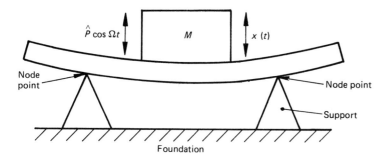

Figure 13.3

13.3.2   Dynamic anti-resonant vibration isolator (DAVI)

As shown in Figure 13.4a, this device consists of a rigid lever carrying a mass, m, and connected to the excited mass, M. Hence if we replace M and m by a dynamically equivalent mass M_e at M (see Figure 13.4b) such that $M_e = M + m(b/a)^2$, then from Section 9.5.1, the amplitude of the vibratory response of M, namely $\hat{X}$, will be

$$\hat{X} = \frac{\hat{P}/K}{1 - \left[1 + \dfrac{m}{M}\left(\dfrac{b}{a}\right)^2\right]f^2}$$

where $f = \Omega/\omega$ and $\omega = \sqrt{(K/M)}$, the undamped natural frequency of the original system. Hence the amplitude of the vibratory response of m, namely $\hat{Y}$, will be equal to $-(b/a)\hat{X}$. Therefore the amplitude of the force transmitted to the foundations, $\hat{FF}$,

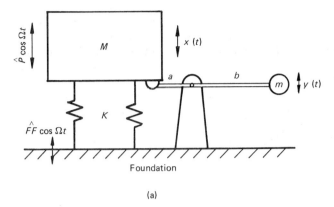

(a)

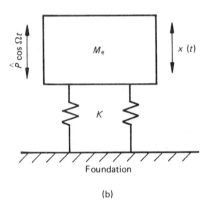

(b)

Figure 13.4

will be the sum of the amplitude of the applied force, and that of the combined inertia force associated with M and m, i.e.

$$\hat{FF} = \hat{P} + \Omega^2 M \hat{X} + \Omega^2 m \hat{Y}$$

giving the transmissibility ratio, $TR = \hat{FF}/\hat{P}$, as

$$TR = \frac{1 - f^2 \dfrac{m}{M} \dfrac{b}{a}\left(1 + \dfrac{b}{a}\right)}{1 - f^2\left[1 + \dfrac{m}{M}\left(\dfrac{b}{a}\right)^2\right]} \tag{13.6}$$

where typical values for the ratios m/M and b/a are 0.01 and 10 respectively. Upon the basis of these ratios, Figure 13.5 is a plot of $|TR|$ to a base of $f(= \Omega/\omega)$. Also, for comparison purposes, superimposed on Figure 13.5 is the corresponding plot for the original system in the absence of the isolator.

From Figure 13.5 we see that for the case where $\Omega = \omega$ the transmissibility is very small when the isolator is incorporated, *and* any small fluctuations in Ω about the value of ω will not give rise to significant increases in transmissibility.

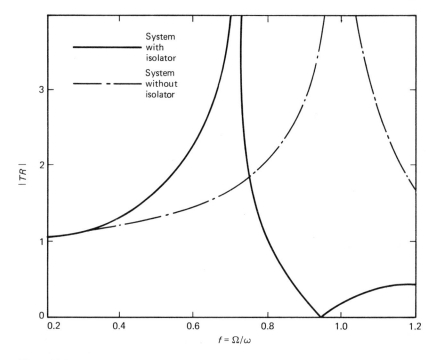

Figure 13.5

13.3.3 Tuned absorber

Let us now turn our attention to absorbers, of which the most common is the *tuned absorber*, which simply consists of a mass, m, attached to the original system by a spring of stiffness k as shown in Figure 13.6.

If we now write down and solve, in the normal fashion, the equations of motion for each of the masses, we deduce that the amplitudes of $x(t)$ and $y(t)$, namely $\hat{X}$ and $\hat{Y}$ respectively, become

$$\hat{X} = \frac{\hat{P}(k - \Omega^2 m)}{(K + k - \Omega^2 M)(k - \Omega^2 m) - k^2}$$

$$\hat{Y} = \frac{k\hat{P}}{(K + k - \Omega^2 M)(k - \Omega^2 m) - k^2}$$

Consequently, as before, the amplitude of the force transmitted to the foundation, $\widehat{FF}$, will be

$$\widehat{FF} = \hat{P} + \Omega^2 M \hat{X} + \Omega^2 m \hat{Y}$$

After simplification, the transmissibility ratio, $TR = \widehat{FF}/\hat{P}$, becomes

$$TR = \frac{1 - \left(\dfrac{\Omega}{\omega_1}\right)^2}{\left[1 - \left(\dfrac{\Omega}{\omega}\right)^2\right]\left[1 - \left(\dfrac{\Omega}{\omega_1}\right)^2\right] - \dfrac{m}{M}\left(\dfrac{\Omega}{\omega}\right)^2} \qquad (13.7)$$

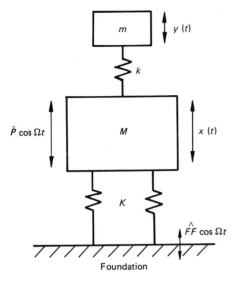

Figure 13.6

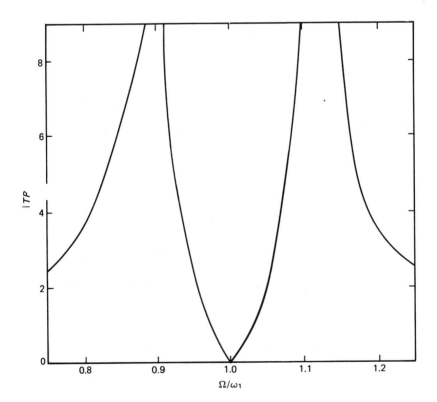

Figure 13.7

where $\omega = \sqrt{(K/M)}$, the undamped natural frequency of the original system, and $\omega_1 = \sqrt{(k/m)}$, the natural frequency of the absorber. From equation 13.7 and Figure 13.7, which shows a typical plot of $|TR|$ to a base of Ω/ω_1 it is readily seen that $TR = 0$ when $\Omega = \omega_1$, i.e. the natural frequency of the absorber is designed to be equal to the frequency of excitation (which may be equal, or close, to the natural frequency of the original system without the absorber). By incorporating this type of absorber, one should however ensure that the excitation frequency does not stray too much from the natural frequency of the absorber since, from Figure 13.7, on either side of this condition lie the two natural frequencies of the complete system i.e. original system plus the absorber. The values of these two natural frequencies can be obtained by equating the denominator of equation 13.7 to zero and solving for the two roots of Ω.

Appendix 1

Standard integrals

$$\int x^n \, dx = \frac{x^{n+1}}{n+1}$$

$$\int \frac{dx}{x} = \ln x$$

$$\int \sqrt{(a+bx)} \, dx = \frac{2}{3b} \sqrt{[(a+bx)^3]}$$

$$\int x\sqrt{(a+bx)} \, dx = \frac{2}{15b^2}(3bx - 2a)\sqrt{[(a+bx)^3]}$$

$$\int \frac{dx}{\sqrt{(a+bx)}} = \frac{2\sqrt{(a+bx)}}{b}$$

$$\int \frac{\sqrt{(a+x)}}{\sqrt{(b-x)}} \, dx = -\sqrt{(a+x)}\sqrt{(b-x)} + (a+b)\sin^{-1}\sqrt{\left(\frac{a+x}{a+b}\right)}$$

$$\int \frac{x \, dx}{a+bx} = \frac{1}{b^2}[a + bx - a\ln(a+bx)]$$

$$\int \frac{x \, dx}{(a+bx)^n} = \frac{(a+bx)^{1-n}}{b^2}\left(\frac{a+bx}{2-n} - \frac{a}{1-n}\right)$$

$$\int \frac{dx}{a+bx^2} = \frac{1}{\sqrt{(ab)}}\tan^{-1}\frac{x\sqrt{(ab)}}{a} \quad \text{or} \quad \frac{1}{\sqrt{(-ab)}}\tanh^{-1}\frac{x\sqrt{(-ab)}}{a}$$

$$\int \frac{x \, dx}{a+bx^2} = \frac{1}{2b}\ln(a+bx^2)$$

$$\int \sqrt{(x^2 \pm a^2)} \, dx = \tfrac{1}{2}\{x\sqrt{(x^2 \pm a^2)} \pm a^2 \ln[x + \sqrt{(x^2 \pm a^2)}]\}$$

$$\int \sqrt{(a^2 - x^2)} \, dx = \tfrac{1}{2}\left[x\sqrt{(a^2 - x^2)} + a^2 \sin^{-1}\frac{x}{a}\right]$$

$$\int x\sqrt{(a^2 - x^2)} \, dx = -\tfrac{1}{3}\sqrt{[(a^2 - x^2)^3]}$$

$$\int x^2 \sqrt{(a^2 - x^2)}\,dx = -\frac{x}{4}\sqrt{[(a^2 - x^2)^3]} + \frac{a^2}{8}\left[x\sqrt{(a^2 - x^2)} + a^2\sin^{-1}\frac{x}{a}\right]$$

$$\int x^3 \sqrt{(a^2 - x^2)}\,dx = -\tfrac{1}{5}(x^2 + \tfrac{2}{3}a^2)\sqrt{[(a^2 - x^2)^3]}$$

$$\int \frac{dx}{\sqrt{(a + bx + cx^2)}} = \frac{1}{\sqrt{c}}\ln\left[\sqrt{(a + bx + cx^2)} + x\sqrt{c} + \frac{b}{2\sqrt{c}}\right] \quad \text{or}$$

$$\frac{-1}{\sqrt{-c}}\sin^{-1}\left[\frac{b + 2cx}{\sqrt{(b^2 - 4ac)}}\right]$$

$$\int \frac{dx}{\sqrt{(x^2 \pm a^2)}} = \ln[x + \sqrt{(x^2 \pm a^2)}]$$

$$\int \frac{dx}{\sqrt{(a^2 - x^2)}} = \sin^{-1}\frac{x}{a}$$

$$\int \frac{x\,dx}{\sqrt{(x^2 - a^2)}} = \sqrt{(x^2 - a^2)}$$

$$\int \frac{x\,dx}{\sqrt{(a^2 \pm x^2)}} = \pm\sqrt{(a^2 \pm x^2)}$$

$$\int x\sqrt{(x^2 \pm a^2)}\,dx = \tfrac{1}{3}\sqrt{[(x^2 \pm a^2)^3]}$$

$$\int x^2 \sqrt{(x^2 \pm a^2)}\,dx = \frac{x}{4}\sqrt{[(x^2 \pm a^2)^3]} \mp \frac{a^2}{8}x\sqrt{(x^2 \pm a^2)} - \frac{a^4}{8}\ln[x + \sqrt{(x^2 \pm a^2)}]$$

$$\int \sin x\,dx = -\cos x$$

$$\int \cos x\,dx = \sin x$$

$$\int \sec x\,dx = \tfrac{1}{2}\ln\frac{1 + \sin x}{1 - \sin x}$$

$$\int \sin^2 x\,dx = \frac{x}{2} - \frac{\sin 2x}{4}$$

$$\int \cos^2 x\,dx = \frac{x}{2} + \frac{\sin 2x}{4}$$

$$\int \sin x \cos x\,dx = \frac{\sin^2 x}{2}$$

$$\int \sinh x\,dx = \cosh x$$

$$\int \cosh x\,dx = \sinh x$$

$$\int \tanh x \, dx = \ln \cosh x$$

$$\int \ln x \, dx = x \ln x - x$$

$$\int \exp(ax) \, dx = \frac{\exp(ax)}{a}$$

$$\int x \exp(ax) \, dx = \frac{\exp(ax)}{a^2}(ax - 1)$$

$$\int \exp(ax) \sin px \, dx = \frac{\exp(a \sin px - p \cos px)}{a^2 + p^2}$$

$$\int \exp(ax) \cos px \, dx = \frac{\exp(a \cos px + p \sin px)}{a^2 + p^2}$$

$$\int \exp(ax) \sin^2 x \, dx = \frac{\exp(ax)}{4 + a^2}\left(a \sin^2 x - \sin 2x + \frac{2}{a}\right)$$

$$\int \exp(ax) \cos^2 x \, dx = \frac{\exp(ax)}{4 + a^2}\left(a \cos^2 x + \sin 2x + \frac{2}{a}\right)$$

$$\int \exp(ax) \sin x \cos x \, dx = \frac{\exp(ax)}{4 + a^2}\left(\frac{a}{2} \sin 2x - \cos 2x\right)$$

$$\int \sin^3 x \, dx = -\frac{\cos x}{3}(2 + \sin^2 x)$$

$$\int \cos^3 x \, dx = \frac{\sin x}{3}(2 + \cos^2 x)$$

$$\int \cos^5 x \, dx = \sin x - \tfrac{2}{3}\sin^3 x + \tfrac{1}{5}\sin^5 x$$

$$\int x \sin x \, dx = \sin x - x \cos x$$

$$\int x \cos x \, dx = \cos x + x \sin x$$

$$\int x^2 \sin x \, dx = 2x \sin x - (x^2 - 2)\cos x$$

$$\int x^2 \cos x \, dx = 2x \cos x + (x^2 - 2)\sin x$$

Basic complex algebra

Consider any two complex factors, $\vec{A}$ and $\vec{B}$, such that

$$\vec{A} = A\underline{/\alpha} = A(\cos\alpha + \mathrm{i}\sin\alpha) = A_r + \mathrm{i}A_i$$
$$\vec{B} = B\underline{/\beta} = B(\cos\beta + \mathrm{i}\sin\beta) = B_r + \mathrm{i}B_i$$

where A_r and B_r are the real components and A_i and B_i the imaginary components of $\vec{A}$ and $\vec{B}$ respectively and i is the complex conjugate, $\sqrt{-1}$.

(1) $\quad \vec{A} + \vec{B} = (A_r + \mathrm{i}\,A_i) + (B_r + \mathrm{i}\,B_i)$
$\qquad\qquad = (A_r + B_r) + \mathrm{i}(A_i + B_i)$

Therefore the addition of $\vec{A}$ and $\vec{B}$ results in a complex factor, $\vec{C}$, such that

$$\vec{C} = C\underline{/\gamma} = C(\cos\gamma + \mathrm{i}\sin\gamma) = C_r + \mathrm{i}C_i$$

where

$$C = \sqrt{[(A_r + B_r)^2 + (A_i + B_i)^2]} = \sqrt{(C_r^2 + C_i^2)}$$

and

$$\gamma = \arctan(C_i/C_r)$$

Similarly,

(2) $\quad \vec{A} - \vec{B} = (A_r - B_r) + \mathrm{i}(A_i - B_i)$
$\qquad\qquad = D = D\underline{/\eta}$

where

$$D = \sqrt{[(A_r - B_r)^2 + (A_i - B_i)^2]}$$

and

$$\eta = \arctan\frac{A_i - B_i}{A_r - B_r}$$

(3) $\quad \vec{A} \times \vec{B} = (A_r + \mathrm{i}\,A_i) \times (B_r + \mathrm{i}\,B_i)$
$\qquad\qquad = (A_r B_r - A_i B_i) + \mathrm{i}(A_r B_i + A_i B_r)$
$\qquad\qquad = \vec{E} = E\underline{/\theta}$

where

$$E = \sqrt{[(A_r B_r - A_i B_i)^2 + (A_r B_i + A_i B_r)^2]}$$

and

$$\theta = \arctan \frac{A_r B_i + A_i B_r}{A_r B_r - A_i B_i}$$

(4) $\vec{A} \div \vec{B} = \dfrac{A_r + i A_i}{B_r + i B_i}$

Now multiplying top and bottom by $(B_r - i B_i)$ gives

$$\vec{A} \div \vec{B} = \frac{(A_r B_r + A_i B_i) + i(A_i B_r - A_r B_i)}{B_r^2 + B_i^2}$$

$$= \vec{F} = F \underline{/\phi}$$

where

$$F = \frac{\sqrt{[(A_r B_r + A_i B_i)^2 + (A_i B_r - A_r B_i)^2]}}{B_r^2 + B_i^2}$$

and

$$\phi = \arctan \frac{A_i B_r - A_r B_i}{A_r B_r + A_i B_i}$$

Basic vector algebra

Consider any two vectors **A** and **B** as shown in Figure A.1 where

$$\mathbf{A} = \mathbf{i}a_x + \mathbf{j}a_y + \mathbf{k}a_z$$
$$\mathbf{B} = \mathbf{i}b_x + \mathbf{j}b_y + \mathbf{k}b_z$$

where **i**, **j** and **k** are unit vectors along the axes OX, OY and OZ respectively.

Vector addition

$$\mathbf{A} + \mathbf{B} = \mathbf{B} + \mathbf{A} = \mathbf{i}(a_x + b_x) + \mathbf{j}(a_y + b_y) + \mathbf{k}(a_z + b_z)$$

Vector subtraction

$$\mathbf{A} - \mathbf{B} = \mathbf{i}(a_x - b_x) + \mathbf{j}(a_y - b_y) + \mathbf{k}(a_z - b_z)$$

and

$$\mathbf{B} - \mathbf{A} = -(\mathbf{A} - \mathbf{B})$$

Vector (or cross) product

$\mathbf{A} \wedge \mathbf{B}$ is defined as a vector having a magnitude $|\mathbf{A}||\mathbf{B}|\sin\theta$, where θ is the angle between the vectors, and $|\mathbf{A}||\mathbf{B}|$ is the product of the magnitude of the vectors.
The vector product, $\mathbf{A} \wedge \mathbf{B}$, can be calculated from the determinant of

$$\begin{vmatrix} \mathbf{i} & \mathbf{j} & \mathbf{k} \\ a_x & a_y & a_z \\ b_x & b_y & b_z \end{vmatrix}$$

i.e.

$$\mathbf{A} \wedge \mathbf{B} = \mathbf{i}(a_y \cdot b_z - b_y \cdot a_z) - \mathbf{j}(a_x \cdot b_z - b_x \cdot a_z) + \mathbf{k}(a_x \cdot b_y - b_x \cdot a_y)$$

Note, however, that $\mathbf{B} \wedge \mathbf{A} \neq \mathbf{A} \wedge \mathbf{B}$, and that *the vector product of two parallel vectors is zero.*

Dot (or scalar) product

$$\mathbf{A} \cdot \mathbf{B} = \mathbf{B} \cdot \mathbf{A} = |\mathbf{A}||\mathbf{B}|\cos\theta$$

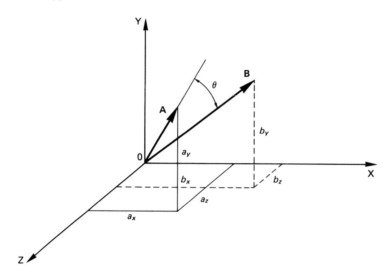

Figure A.1

Therefore

$$\mathbf{A} \cdot \mathbf{B} = (\mathbf{i}a_x + \mathbf{j}a_y + \mathbf{k}a_z) \cdot (\mathbf{i}b_x + \mathbf{j}b_y + \mathbf{k}b_z)$$

and since

$$\mathbf{i} \cdot \mathbf{i} = \mathbf{j} \cdot \mathbf{j} = \mathbf{k} \cdot \mathbf{k} = 1$$

and

$$\mathbf{i} \cdot \mathbf{j} = \mathbf{j} \cdot \mathbf{i} = \mathbf{k} \cdot \mathbf{i} = \mathbf{i} \cdot \mathbf{k} = \mathbf{j} \cdot \mathbf{k} = \mathbf{k} \cdot \mathbf{j} = 0$$

then

$$\mathbf{A} \cdot \mathbf{B} = a_x \cdot b_x + a_y \cdot b_y + a_z \cdot b_z$$

Note that the *dot product of two perpendicular vectors is zero.*

Index